THE ASSEMBLAGE BRAIN

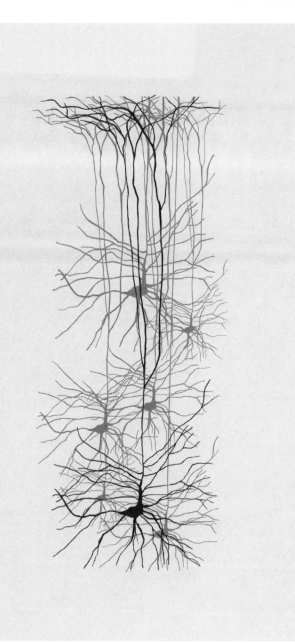

Illustration by Dorota Piekorz.

THE ASSEMBLAGE BRAIN

Sense Making in Neuroculture

Tony D. Sampson

University of Minnesota Press
Minneapolis · London

Portions of part II were published in "Tarde's Phantom Takes a Deadly Line of Flight: From Obama Girl to the Assassination of Bin Laden," *Distinktion: Scandinavian Journal of Social Theory* 13, no. 3 (2012), special issue, "Operations of the Global: Explorations of Connectivity."

Published by the University of Minnesota Press
111 Third Avenue South, Suite 290
Minneapolis, MN 55401-2520
http://www.upress.umn.edu

Printed in the United States of America on acid-free paper

The University of Minnesota is an equal-opportunity educator and employer.

22 21 20 19 18 17   10 9 8 7 6 5 4 3 2 1

Library of Congress Cataloging-in-Publication Data
Names: Sampson, Tony D., author.
Title: The assemblage brain : sense making in neuroculture / Tony D. Sampson.
Description: Minneapolis : University of Minnesota Press, 2017. |
Includes bibliographical references and index.
Identifiers: LCCN 2016015340 | ISBN 978-1-5179-0116-5 (hc) |
ISBN 978-1-5179-0117-2 (pb)
Subjects: LCSH: Brain—Anatomy. | Neurosciences.
Classification: LCC QM455 .S16 2017 | DDC 612.8/2—dc23
LC record available at https://lccn.loc.gov/2016015340

Dedicated to John Stanley Sampson (1960–1984)

Contents

Introduction

THE IMMINENT SHADOW OF NEUROCULTURE

This book develops a radical critical theory concerned with the expansion of a neurocentric world into all corners of cultural, social, political, and economic life—what I will tentatively call here *neuroculture*. Neuroculture is, evidently, nothing new. Its modern origins can be traced back to Santiago Ramón y Cajal's anatomical drawings of brain cell structures and the subsequent neuron doctrine established in the latter part of the 1800s. Nonetheless, there has been a pronounced acceleration of the influence of neuroculture in recent years wherein Arthur Kroker notes, "Everything seems to increasingly operate . . . under the ideological sign of the neuro."[1] Neuroculture is indeed an expression of the offshoots of the neurosciences manifested in what Edmund T. Rolls catalogs as neurosociality, neuroreason, neurophilosophy, neuroaesthetics, neuroeconomics, neuroethics, neuropsychiatry, neuroreligion, neuroaffect, and neuropolitics.[2] The ubiquity of neuroscience is, like this, usurping the power of the other big sciences, including genetics and computer science, in terms of explanatory discursive power, attracting funding, and gaining global media exposure.[3] Certainly, more than a quarter of a century after former U.S. president G. W. Bush declared that a new era of discovery was dawning in brain science research, billions of dollars have been invested in diverse neuroscience-related research collaborations between the state, the university, and the business enterprise.[4]

The expansion of neuroculture has been energized, to a great extent, by the advent of an array of neurotechnologies, including neuroimaging, noninvasive brainwave measurements, neurodevice implants, and neuropharmacological interventions. These new technologies, and the experiments they help initiate, are, according to numerous popular accounts, supposedly unlocking the secrets of brain functionality, including those assumed to be responsible for particular behaviors and making conscious experience possible. Neurotechnologies have certainly helped to extend the reach of the brain sciences into many nonmedical arenas, and for that reason, they have presented new ethical challenges concerning the potential to intervene directly in brain processes relating to, for example, educational performance, labor efficiency analysis, consumption, and economic decision making. There are indeed concerns to be expressed over the extent to which the brain is objectified by neurotechnology, on one hand, and the role neuroculture plays in multiple processes of subjectification, on the other. In the first instance, neuroculture has inspired managerial efforts to put the brain, and, more specifically, the neuron, to work in more efficient, and implicit, ways. In the second instance, it is my contention in this book that a concoction of neurotechnology, neurospeculation (vis-à-vis how subjectivity emerges), and neurologically inspired business enterprise intervenes in processes of subject making, that is to say, how *subjectivity in the making* occurs in accordance and consistent with particular cultural, political, and economic conditions under capitalism. This expression of neuroculture is what I will call in this book *neurocapitalism*.

Neuroculture manifests itself in a number of ways. To begin with, we can point to a more generalized process of subjectification wherein neuroimaging techniques have resulted in highly publicized speculation concerning such things as gender difference, political preference, and economic behavior, all supported by what appear to be fairly crude correlations made between certain performances, beliefs, elusive and narrow psychological categories, and regionalized cerebral processes, such as blood flow, brainwave frequency, and neurochemical transmission. In the United States, for example, neuroscientists have claimed that MRI scans can help to reveal brain differences between Republicans and Democrats according to neural activity in areas believed to be linked with social connectedness.[5] Beyond its clinical remit, then, neuroimaging-based speculation not only endeavors to objectify the neuron but also informs ideas about what constitutes emerging subjectivity, which have consequently found a practical application in political psychology, economics, education, legal policy, social theory, aesthetics, ethics, and philosophy.[6]

The first part of this book (chapters 2 and 3) focuses specifically on the role neuroculture plays in informing significant developments in computer work, marketing, and the neuropharmaceutical control of inattentiveness in the schoolroom. All three of these developments are seemingly propelled by neurologically inspired managerial drives for more efficient and seamless human interactions within digitized workplaces, sites of consumption, and educational institutions. To begin with, the theories and practices of human–computer interaction (HCI) have shifted away from the engineering metaphors of a paradigm shaped by cognitive psychology toward the so-called *emotional turn* in the brain sciences, suggesting a new paradigm of research into the situated experiences of computer work.[7] Human interactions previously at the margin of the cognitive paradigm, like emotions, affect, feelings, and fun, become central to a new metaphor of interaction grasped as "phenomenologically situated."[8] Likewise, new commercial collaborations between the neurosciences and business enterprise have inspired novel marketing techniques that can, it is claimed, tap into nonconscious, emotional brain processes that influence purchase intent. Of course, the claims of the neuromarketer have not, unsurprisingly, attracted much skepticism;[9] however, these marketing practices are endemic to an economic regime that strives to situate an increasingly docile consumer-subject managed according to channeled attention, primed emotional engagement, and visceral affective stirrings. So although many of the practices of neuromarketing might not survive rigorous scientific scrutiny, they continue to legitimize numerous attempts to subliminally market to consumers by way of appealing to implicit brain functioning, however vaguely understood. Similarly, the pharmaceutical industry's complicity in the sharp rise of psychostimulants used to *care for* attention-deficit hyperactivity disorder (ADHD) in the classroom draws further attention to the control of behavior through direct interventions into brain functioning. These medicated care systems, aimed at normalizing the problem of inattention (and hyperactivity), are, like neuromarketing, equally founded on speculative brain science. Following on from controversial neurotypical diagnoses, psychostimulants simultaneously stimulate cognitive processes and sedate behavior, producing a certain kind of neurosubject in the classroom—one who is, at once, attentive and subdued.[10]

Although these examples of neurospeculation need to be treated cautiously, neuroculture's role in subjectification processes should not be ignored or underestimated. Central to my approach in this book is a recognition that what might now appear to be a period of epistemological ignorance concerning the possibilities of brain science should not get in the way of ontological concerns about neuroculture. Indeed, by taking neuroculture seriously, this book intends to draw attention to two significant problems that appear to be central to the thorny relation established between science and critical theory. The first concerns how critical theory should engage with the concurrence of neuroculture and capitalism. The contribution in the first part of this book is resolutely focused on grasping the political, social, and cultural purchase of neuroscientifically inspired modes of market capitalism. My aim is to meet head on a rapid territorialization of everyday life by neuroculture, spreading out from early studies of brain anatomy and function to more recent interventions into brain processes currently unfolding in the transformative cultural circuits of capitalism.[11] This is a circuitry wherein neuroculture becomes increasingly mapped onto regimes of free market capitalism as evidenced in digital labor, marketing power, and big pharma. To be sure, as it becomes more and more rehearsed, neuroculture develops into an integral part of these circuits, casting an often murky shadow across everything it encounters. Moreover, when exposed to a certain kind of *neurorationalism,* the imminent shadow of neuroculture can become a deterministic and functionalizing mechanism, ever more apparent in the application of brain imaging in neurophilosophy and neuroaesthetics, wherein conceptual thinking and aesthetic practices can all too often be reduced to a kind of crude locationist *neurophrenology.* A major concern is that such a deterministic shadow might lead to a hideous unification of the concepts of the philosopher and the sensations of the artist under the ominous flag of rational science.

Relatedly, and of equal concern perhaps, is the aforementioned pervasiveness of the emotional turn in the brain sciences, which has, to some extent, challenged the doctrine of the rationalist approach in neuroscience. This innovative focus on how feelings, affects, and emotions influence cognition has ignited the curiosity of the business enterprise, keen to tap into nonconscious brain processes in the workplace and the shopping mall. It is, as such, important to note how the work of Antonio Damasio and Joseph LeDoux, for example, has moved to the forefront of a far-reaching fascination with the emotional brain, evident in, for example, interaction design and marketing but also cultural studies, social theory, and philosophy alike. There are those in the humanities who are perhaps understandably cautious about the seemingly indiscriminate poaching of the brain sciences in these latter disciplines, particularly those approaches that throw into question phenomenological and sociological interpretations of subjectivity as self-contained embodiment.[12] There is certainly justification for wider anxieties too. The shadow cast by the merging forces of the neurosciences and capitalism is going to be, most probably, a gloomy, dystopian affair, requiring a heightened criticality. It is, however, the objective in the second part of this book (chapters 4 and 5) to find at least some light in among the contested shadows of neuroculture. There is room, I contend, for a critical theory that routes around old predispositions resulting from too much theoretical distance between philosophy and science and gets under the skin of neuroculture to tease out something of value from critically engaging with the brain sciences at close quarters.

The second problem approached in this book concerns the poverty of the culture–biology fissure. My aim here is to push the limits of this exhausted perspective, which is, I contend, caught up in an unremitting theoretical logjam in the social sciences and humanities. This standstill is probably at its most visible in recent disputes between representational and nonrepresentational theorists, where a thick line has been drawn between discourse analysis and affect studies.[13] However, while acknowledging that, by engaging with neuroculture, there is always a risk of ushering in a deeply unsatisfactory and naive biological deterministic viewpoint of culture, through notions of brain plasticity, for example, it is equally important to challenge the mainstay of archaic approaches to cultural theory, which absolutely reject any kind of biological perspective on culture in favor of a dogmatic socially constructed worldview. It is my intention that this study of neuroculture should move beyond the culture–biology impasse and, as a consequence, pose new questions concerning the emergence of subjectivity in the making in which biological and cultural distinctions are collapsed into each other. Indeed, many of the assumptions concerning the emergence of subjectivity, and how, sequentially, the brain–somatic relation becomes related to the world, are already influenced, in one way or another, by the shadow of neuroculture. To be sure, although nonrepresentational theory may well be accused of haphazardly poaching from scientific knowledge concerning affective brain states, representational theorists similarly remain loyal to an equally unstable cognitive paradigm that believes in a mind that thinks in images. This is a locationist model of the brain that has special access—it would seem, to a cerebral photographic album. The approach developed in this book is instead focused on what the antilocationist Henri Bergson refers to as the insensible degrees between the *unextended* affective state of things and the ideas and images that represent and occupy them in space.[14]

Theories of subjectivity in the humanities have tended to depend on phenomenological ready-made concepts, bound together by fixed self–other identities, personalities, personal experiences, opinions, and perceptions. The intention here is to route around these ready-mades by initially exploring theories of emergent subjectivity debated in the neurosciences before going on to develop a concept of the *brain-becoming-subject* (or protosubjectivity) that I will call the *assemblage brain*. The aim is not to uncover the mechanisms that determine the experience of conscious awareness[15] but to politically grasp affective realms of sense making beyond the limits of locationist doctrines in philosophy and science. Indeed, although the substitution of the Cartesian transcendence of mind over matter for the synaptic brain of Damasio and LeDoux's emotional brain opens up a valuable new materialist perspective, it does not go far enough in terms of shaking off the locationist tendencies apparent in the engineering metaphors of the cognitive paradigm. Moreover, the computer mind does not allow us to grasp the brain-somatic encounter with sensory environments beyond a locationist image of thought, that is to say, a perception of the world established in an inner brain state looking outward. Indeed, the concept of the assemblage brain provides an antilocationist image of thought intended to stretch the notion of neurodiversity beyond the cellular walls of the neuron to an ecology of sense-making environments and encounters.

THE ASSEMBLAGE BRAIN

The assemblage brain intervenes and routes around preceding models of perception in the brain sciences, from the behaviorist black box to the cognitivist computer mind and, more recently, the emotional brain, which have all maintained, by degree, a locationist tendency that grasps subjective experience as an internal representation containing a model of the external world. However, although intervening in these images of thought, it is important not to mistake the assemblage brain for the idealist's critical confrontation with science. In many ways, the critical distance established between the hardwired inner model of self (biologically understood by the neurosciences) and an understanding in the humanities of a cultural world of human ideas has led us to a wasteful theoretical impasse. Any slippage, we are told, toward a posthuman (or nonhuman) paradigm in the humanities threatens to open the door to the neurosciences and will, it would seem, lead to the ruination of human ideas.[16] Biological and idealist images of thought are accordingly challenged in this book by way of the application of an assemblage theory of sense making: an alternative image of thought.

This assemblage brain follows a trajectory beginning with the panpsychism apparent in the work of Henri Bergson, Gabriel Tarde, Gilles Deleuze and Félix Guattari, and elsewhere in the work of Alfred N. Whitehead, Raymond Ruyer, and Gilbert Simondon; that is to say, what becomes brain matter does not need to begin in a location or with a neuron at all. The proliferation of cells in the embryo, for instance, is without neuronal knowledge, yet clearly these cells engage in a kind of relational sense making.[17] There are indeed nonhumans to consider, such as the lamprey, the sponge, and the slime mold, who make sense of the world without access to neurons.[18] Moreover, though, what becomes brain need not necessarily begin with a cellular organism at all. The relational forces of nonorganic matter, including atoms and molecules, are sense makers without neurons. As Deleuze and Guattari argue, "not every organism has a brain, and not all life is organic, but everywhere there are forces that constitute micro-brains, or an inorganic life of things."[19] These microbrains are a generalized force of *contemplation* found in "rocks and woods, animals and men . . . even our actions and needs."[20] The assemblage brain needs to be traced all the way to these microbrains of inorganic life—a self-awareness *without knowledge* that anticipates the world, not through philosophical concepts in the cognitive sense, but via the vibratory sensations of all matter. Every*thing* is, like this, potentially *becoming brain*.

There is clearly a considerable mismatch between the territorializing forces of cognitive neuroscience and a concept of nomadic thought borrowed from Deleuze and Guattari wherein the sense making of microbrains aspires to be at its most deterritorialized. Indeed, if we are ever to arrive at anything resembling a state of endless potential, or, as I suspect, a much more modest version of it, we must begin by establishing the brain's relation to chaos, grasping it, as Deleuze and Guattari do, as a junction, or point of exchange, between infinite possibilities and the brain-becoming-subject. Then, nomadic thought will need to break out of the territorializations of the locationist's brain models that incarcerate sense making in the human cranium. Significantly, this endeavor will require a questioning of both Deleuze's tendency to lean on the neurosciences and a reappraisal of Deleuze and Guattari's notion that it is not the person or mind but the material brain that *thinks*.[21] To do this, we need to begin within the shadows of neuroculture.

Following a reading of Deleuze and Guattari's methodology in *What Is Philosophy?* in chapter 1, this book introduces a series of conceptual personae, partial observers, and aesthetic figures (e.g., Cajal's Mira-honda, Gramsci's nonconformist brain, the lab rat, Huxley's soma, and Sholim's thumb) intended to help explore the limits and potential of neuroculture through *interferences*. These virtual sense makers and subjectless forces move cautiously through the terrain of the brain sciences, avoiding the locationist's tendency to theorize subjectivity from the inside looking out. Indeed, the antilocationist stance I develop in the second part of the book coincides, to some extent, with what Deleuze and Guattari call a *nonphilosophy*: a philosophy that does not explain but produces.[22] Nonphilosophy is grasped here as a kind of production of a heterogeneous encounter that *thinks* only on the outside. It becomes, as follows, a *nonlocalized* nomadic intervention into neurocentric accounts that looks further than the human example of thought located inside the brain. Sense making is not stuck in the human skull, as such; it is distributed throughout brain–somatic chemistries, intersubjective crowds, digital networks, nonhuman (and posthuman) sense makers, and beyond, into the cosmos of relational assemblages that bring organic and nonorganic matter together.

CRITICAL INTERFERENCE AS METHOD

The assemblage brain concept stems from a critical interference between philosophy, science, and art, which not only provides an alternative to the notion of critical distance but also differs from the conventional analytical philosophical approach to the brain sciences. Although the method deployed similarly draws attention to issues that would interest an analytical philosopher, such as an emergence theory of sense making, it does so in ways that are intended to produce novel, experimental, and significantly political concepts. This is an approach that may (or may not) sit well with those readers dedicated to analytical methods, because it is not driven by the logical testing of concepts; rather, the assemblage brain draws on the interferences between philosophy, science, and art to think through the boundless potential of concepts and to probe, at the same time, the political implications of neuroculture. Importantly, the critical interference provides an intervention into critical distances and analytic philosophies at a time when the imminent colonization of brain processes by capitalism demands that critical theory look for a new dialogue between philosophy and science. This is a political mode of interference that in many ways situates philosophy, science, and art in the rhythmic interferences of a rampant neurocapitalism. Beyond the remit of this book, then, the critical interference may well find a place in the cross-disciplinary spaces of the university already exposed to these political rhythmic interferences. This is a critical theory that might resonate with science through art practice, for example.

The first chapter, "Interferences: Philosophy, Science, Art, and Capitalism," therefore sets out a methodological basis for such an encounter with neuroculture. It borrows from Deleuze and Guattari's perhaps most controversial book together to explore the potential for a method that produces interferences between the three discrete planes of enunciation in the sciences, philosophy, and art. Indeed, throughout the book, the trajectories of these three planes are forced to traverse each other. Like this, the critical interference is grasped as a way to constructively (or destructively) displace the discreteness of each plane. Following from Tarde's social theory, the interference is also understood here as a *site of invention* intended to produce short circuits, or social crossroads, breaking away from the constraints of the repetition of the same.[23] For example, there are problems posed by various rationalist neuroscientific incursions into philosophy and art that this methodology needs to be critically attuned to, and possibly antagonistic toward, but the interference is *not* an attempt to pit a socially constructed brain against a biologically constituted brain. Again, the book aims to collapse the distinction between the two.

Rather than oppose neuroculture to some higher order of relativist or idealist philosophy, my intention here is primarily to grasp the interferences between science, philosophy, and art as an expression of the brain's encounter with chaos. This method will require a theoretical framework that juxtaposes scientific functions, philosophical concepts, and the sensations of art practice. Like this, the limiting effects of scientific propositions concerning functions of knowledge, which might otherwise turn philosophers and artists into intolerable rational subjects, need to be forced into contact with the endless and open possibilities of concepts and sensations. There is indeed a significant need for a radical critical theory that interrogates two kinds of neuroculture. First, there is a neurorationalism that proposes, for example, a totalizing political project based on a "rational social contract."[24] This is an extension of Thomas Hobbes's political project, which takes a particular version of human rationality as its starting point and end goal. In terms of art, too, neurorationalism introduces a conservative neuroaesthetics that claims to be able to discern between an objective aesthetic of beauty and the "dubious" *imposters* of conceptual art.[25] Marcel Duchamp is, like this, presented as an absurd figure who, as any child in an art gallery can apparently see, parades himself in the emperor's new clothes.[26] As the search for correlates between neuronal matter and mental states in the localizing techniques of neuroimaging research evidences, there is a tendency in these rationalist enunciations of neuroculture to turn concepts and sensations into coordinates on a grid, trapping them inside a function of knowledge that can be intervened into, adapted, and made more pliable to further intercessions. To some extent, then, neuromodeling of this kind opens up the neuron to experimentation in new, nonmedical arenas like the workplace and shopping mall. It is an endeavor to produce a flexible brain, which, as Catherine Malabou contends, coincides with capitalism, that is to say, what Malabou describes as an *ideological avatar* that obscures what it really means for a brain to be free.[27] Second, there is the emotional turn in brain science to contend with. On one hand, the potential of a material brain, in which affects are enmeshed in concepts, opens up sense making to the relational capacities of the sensory environments brains inhabit. Certainly the plasticity of such relations has led philosophers like Malabou to consider the model of the synaptic self as a brain, which, although unware, is already free.[28] On the other hand, though, there is a need to grasp that the sensory environment or affective atmospheres, in which brains are becoming subject, are increasingly being exploited by what has been called affective capitalism or what I will specifically term here *neurocapitalism* so as to draw attention to its progressively neurocentric nature. Indeed, the brain's plastic relation to sensory environments is ever more mediated through the *ersatz experiences* of the marketer in which the brain is subsumed in preprimed affective atmospheres that route around cognition and appeal directly to mood.

TWO QUESTIONS FOR NEUROCULTURE

Following the first chapter's introduction to critical interferences, the book moves on to address two central questions. The first asks what can be done to a brain and follows the trajectory of neuroculture from its initial focus on anatomy and function to the current wave of neuro-technological interventions. In short, this first part of the book ad-dresses relations of neuropower established between the potential for neuroscientific control of digital labor and consumption, on one hand, and marketing and big pharma, on the other. Chapter 2, "Neurolabor: Digital Work and Consumption," begins by exploring the point at which the cultural circuits of neurocapitalism converge in HCI, user experience (UX), and ubiquitous computing. It looks specifically at a concept of neurolabor informed by the continuities and discontinui-ties of a shift from the management of material labor in the Taylorist factory to post-Taylorist immaterial labor. These capricious and fluctu-ating shifts are traced through three paradigms of HCI, each illustrat-ing the fundamental positioning, repositioning, and exploitation of brain–somatic relations in managerial strategies of control intended to (1) eradicate nonconformity and (2) produce more efficient digital labor. The chapter argues that the current paradigm of HCI research marks a convergence between digital work and consumption situated in a neurologically inspired economy based on ersatz experiences—a deepening of a technological unconscious and the control of sensory environments achieved by way of the analysis of efficient emotional performances with, for example, gamified interactions with smart, social, and pervasive digital media.

Chapter 3, "Control and Dystopia," makes a comparison between the emotionally engineered control systems imagined by Aldous Huxley in the early 1930s and the dystopian expressions of neuroculture apparent in neuromarketing, neurotechnology, and the psychostimulants used to *treat* ADHD today. Here the parallels between Huxley's hypnopedic rhythms and current brain wave technologies, on one hand, and the Brave New Worlders' use of soma and the rise in the use of drugs like Ritalin, on the other, are discussed. The chapter also introduces the idea that through the interventions of neuromarketing and neuropharmacy, capitalism increasingly taps into rhythmic, affective brain–somatic states with the intention of, it seems, entraining brain frequencies to the quickening cadence of postindustrial life.

The second question asks not what can be done to a brain but what a brain can do. Aside from the palpable Deleuzian–Spinozian overtures regarding the capacity of a body to affect and be affected by its environment, what initially becomes key to grasping this second question is that we might not realize, as Malabou argues, the full potential of the brain–somatic relation to be free. There is a need therefore to become

> acquainted with the results of current discoveries in the neurosciences in order to have an immediate, daily experience of the neuronal form of political and social functioning, a form that today deeply coincides with the current face of capitalism.[29]

To be sure, like Tarde's social somnambulist, the brain we find in times of neurocapitalism seems to be in a constant mode of sleepwalking toward its own repression. As follows, Malabou's notion of neuronal liberation is intriguingly grasped because it provides an alternative malleable prototype of subjectivity that might awaken from the rigidity of the cybernetic model of the mind—a model already put to work in the efficiency analysis of digital labor. However, arguably, this notion of brain freedom needs to be radically expanded to comprise more inclusive assemblages of sense making than those provided by the model of neuroplasticity. Certainly, if the brain is to wake up from this Tardean *dream of action,* it needs to realize its capacity to reach out to what Tarde regarded as a universal social condition not restricted to the inner world of human subjectivity but rather *out there* in the infinite possibilities of imitative social encounters.

Chapter 4, "Sense Making and Assemblages," begins by evaluating three alternative brains that each challenge the cognitive paradigm that has dominated discussion on individual and collective sense making. First, Malabou's favored *synaptic self* is explored before the discussion moves on, via Bergson's antilocationist position, to entertain a mereological challenge to this manifestation of the neuron doctrine and its central claim that the "sum total of who we are" is located *inside* the brain. At first glance, this second brain—a delocalized reworking of the whole–part relation expressed in Wittgenstein's *systematic brain*— provides an inclusivity of body parts engaged in sense making other than the brain.[30] Nevertheless, we find that the liberation of mouths, eyes, fingers, and thumbs from the neurocentric worldview becomes equally limited to a *whole system* that miraculously transcends the interaction between its parts. In contrast, the parts of a third assemblage brain never become a sum total or systemic whole but are in a continuous diachronic relation with each other, ceaselessly becoming lost and regained, attached and detached, and engaged in relations of exteriority. Indeed, chapter 4 concludes by establishing a link between an assemblage theory of emerging protosubjectivity and Tarde's panpsychism, the latter of which conceives of sense-making processes of desire and belief not beholden to organic matter alone.

Finally, chapter 5, "Relationality, Care, and the Rhythmic Brain," continues to intervene in the locationist tendencies of neuroculture by opening up the brain to outside forces. Here the forces of exteriority Deleuze introduces in *The Fold* are contrasted with the neurophilosopher Thomas Metzinger's extension of the cybernetic brain (and evocation of Plato's cave) to describe a brain that hallucinates itself, and the world around it, so as to cope with the evolutionary pressures the organism is subjected to; that is to say, Metzinger's sense of self is an illusion of the outside retained in a virtual simulator located in the inner workings of the brain.[31] Significantly, *The Fold* presents a very different kind of sense-making relation to the world wherein the inside is nothing more than a fold of the outside. Indeed, looking beyond the shadowy caves and tunnels of Metzinger's brain, this final chapter identifies the folds of sense making in the Tardean imitative encounters Christian Borch well describes as processes of rhythmic subjectification in urban environments.[32] This is a multiple process of environmental subjectification John Protevi similarly locates in Bruce E. Wexler's *Brain and Culture*.[33] Herein, Protevi's notion of *radical relationality* articulates a brain-becoming-subject that does not need to become situated in an oppositional relation between inner and outer worlds. Radical relationality introduces a politics of affect that can help critical theorists better grasp, and potentially contest, the dystopian tendencies in neuroculture that target brain waves to normalize and control an otherwise neurodiversity of brain rhythms. Moreover, the chapter concludes by contending that a better understanding of the significant role caregivers play in mediating imitative encounters with sensory environments might lead to alternative care systems to those that locate the problem of ADHD, for example, on the inside.

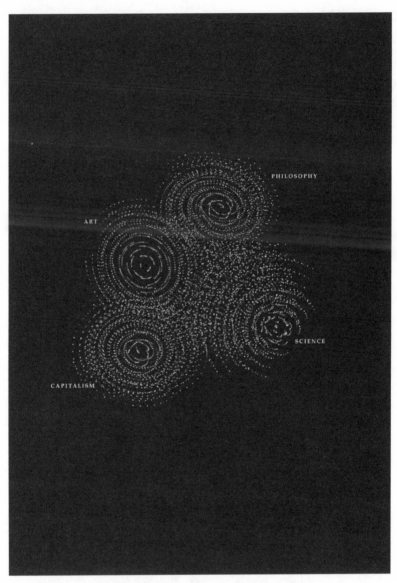

Interference as method. Illustration by Francesco Tacchini.

1

Interferences: Philosophy, Science, Art, and Capitalism

THE BRAIN IS AN OCEAN

In the age of the neuroimage, the metaphorical notion of a brain that is comparable to a software package running on a computer system is, it is claimed, untenable. Figuratively speaking, it is better to think of the brain as an ocean rather than as an information system.[1] This is a highly compelling alternative metaphor indeed. In place of the engineering-inspired architectures of inputs and outputs, short-term and long-term memory storage, and representational encoding, the ocean brain presents an encounter with chaotic forces. In lieu of silicon circuits, the ocean brain has gray, wrinkled surfaces through which complex and fluid currents flow. These smooth, undulating surfaces are further shaped by waves stirred into action by storms as chaotic as any planetary weather system.[2] To fully grasp sense making, it is not computer science that we need but an oceanography of the brain.

Perhaps the chaos of the ocean brain is more than a mere metaphor. Certainly, unlike the highly abstracted analogies of the cognitive brain, the waves of the ocean brain point to a more definable objective reality in which the brain becomes related to the world. To be sure, although there is no visible trace of information or representation in the neuroimage, the affects of environmental stimuli on brainwave frequency, like those produced by sound waves, can be identified in electroencephalograph (EEG) research. Although not exactly revealing cause-and-effect mechanisms, the crashing waves of sensory stimuli evidently influence the brain's rhythmic frequencies, which, some researchers contend, relate in some way to entrainments, vigilance, and mood.[3] Moreover, unlike the information channels of the cognitive brain, which connect the inner world of representation to the external world of environmental stimuli, wave forms are constructively (or destructively) displaced and distributed as intensities on an undulating surface. What relates the waves of environmental stimuli to the waves of the brain is the *interference*.

The ocean brain is not a selfdom trapped inside a black box, a computer, or, for that matter, any other object explained by an engineering metaphor. In contrast to a brain that processes and stores information internally, the interferences of the ocean brain suggest a kind of physics that only *knows* an outside. To be sure, there are no closed subjects relating to solid external objects, no continuous surfaces or straight lines running between internal and external realities: *only waves*. The brain is, like this, implicated in processes of subjectivity making understood as open to the rhythmic disturbances of the outside world. The interference is indeed the brain's encounter with chaotic external forces. In physics, for example, interferences can be looked at in very specific ways. They are, as Arkady Plotnitsky points out, "the addition or superposition of two or more waves that results in new wave patterns."[4] The interference is, as such,

> the interaction of waves that are correlated or coherent with each other, either because they come from the same source or because they have the same, or nearly the same, frequencies. If one has, as in music, two complex multi-periodic wave systems comprised of multiple harmonics which can interfere with each other, we can see a very complex resulting pattern of amplitudes in which the initial amplitudes are added and subtracted in multiple ways.[5]

As follows, the physics of the interference can be mapped to binaural waves in rhythmic frequency entrainments associated with affective brain states, for example.[6]

A broader concept of interferences can also admit the critical theorist to a speculative understanding of the relations established between different brains. For that reason, this chapter borrows from the idea of the interference introduced at the end of Deleuze and Guattari's *What Is Philosophy?* and considers its worth as a means by which to disrupt the seemingly discrete divisions that separate philosophy, science, and art (the *three planes of enunciation*). In short, three kinds of interference (*extrinsic, intrinsic,* and *nonlocalized*) become part of an experimental method intended to open up the many points at which the inventions of philosophers, scientists, and artists (concepts, functions, and sensations) meet. Interferences can enable the critical theorist to slide between planes and enter into the chaos that traverses disciplinary boundaries. Ultimately, these planes become indistinct from each other, culminating in a kind of *nonphilosophy, nonscience,* and *non-art.*[7]

To begin with, in this chapter, I look briefly at Deleuze and Guattari's tendency to lean on the brain sciences and ask what kind of brain (or brains) we subsequently encounter in their philosophy. There is not a straightforward answer to this question, because between the publication of *A Thousand Plateaus* in 1980 and *What Is Philosophy?* in 1991, we encounter at least two kinds of brains, both of which need to be clearly delineated before the purchase of the interference can be fully grasped. The chapter then moves on to unpick some of the key concepts that appear in *What Is Philosophy?* First, a clear distinction is made between philosophical concepts, scientific functions, and the sensations of art, before proceeding to think through how each can be brought to life, observed, and felt by the respective deployment of a series of tools, including *giants, conceptual personae, partial observers,* and *aesthetic figures.* Indeed, it is these deployments (in unequal measure) that enable us to slip from one plane to the next, between concept, function, and sensation. Finally, the chapter returns to the three planes of enunciation to discuss their problematical discreteness from each other.

N IS FOR NEUROLOGY

Before we can grasp the significance of the interference as a methodological approach to neuroculture, it is important to briefly stress just how fundamental the brain is to Deleuze and Guattari's ontological worldview and ask what kind of neurophilosophers this fascination with neurology makes them. Focusing initially on Deleuze, we can see how his self-taught fascination with the neurosciences divulges a long-standing commitment to the future of the molecular biology of the brain over any other science, including information and communication theory[8]—as he makes clear, his "inspiration doesn't come from computers but from the microbiology of the brain."[9] Moreover, his long-held suspicions concerning psychoanalysis seem to have also been prompted, to some extent, by the fissure Freud opened up between psychiatry and the sciences of the brain early in his career.[10] For instance, in 1985, Deleuze approaches a philosophy of cinema by telling Gilbert Cabasso and Fabrice Revault d'Allonnes that they should not look for principles in psychoanalysis (or linguistics) but instead look to the biology of the brain for answers.[11] This is because, he contends, the latter "doesn't have the drawback, like the other two disciplines, of applying ready-made concepts."[12] He continues in 1988 to tell Raymond Bellour and François Ewald that there is a special relation between philosophy and neurology, going back to the associationists, to Arthur Schopenhauer and Henri Bergson.[13]

The brains we encounter in *A Thousand Plateaus* and *What Is Philosophy?* are nonetheless seemingly very different propositions. The rhizome brain is indeed the outcome of interferences between philosophy and early neuroscience. Its discontinuities are, it would seem, derived from Santiago Ramón y Cajal's nineteenth-century anatomy of the neuron network and, in particular, his challenge to the reticularist's notion that the brain is a continuous Golgi network (see the introduction to part I for more detail). Significantly, though, unlike the neuron doctrine, the rhizome brain does not become a mere location in which thought functions. It is not constrained to an algorithmic relation between neurons. Instead, Deleuze and Guattari grasp the rhizome brain as a mode of limitlessness and relational nomadic thought opposed to logic, communication, and arborescent evolutionary thinking. Some eight years after *A Thousand Plateaus,* Deleuze continued to grasp the brain's organization as "more like grass than a tree"—a probabilistic and uncertain system, defined not by the starting and finishing points of thought but by the capriciousness of the journey itself:

> It's not that our thinking starts from what we know about the brain but that any new thought traces uncharted channels directly through its matter, twisting, folding, fissuring it.... New connections, new pathways, new synapses, that's what philosophy calls into play as it creates concepts, but this whole image is something of which the biology of the brain, in its own way, is discovering an objective material likeness, or the material working.[14]

In contrast, the chaos brain, as initially encountered in *What Is Philosophy?*, seems to be at odds with the limitlessness of the rhizome. There is a surprising discreteness established between the brains of philosophers, scientists, and artists that is clearly inconsistent in many ways with the rhizomatic mixtures of *A Thousand Plateaus*. It is, nonetheless, a brain that is entirely consistent with ongoing confrontations between philosophy, science, and art, wherein, it must be said, the warring factions are still busy today drawing up their respective battle lines. On one hand, some neuroscientists claim that philosophers and artists have, by and large, failed to answer the big questions concerning, for example, what constitutes subjective human experience or the essence of beauty. The philosopher and artist may indeed learn a thing or two from the neurosciences, they claim.[15] Given advances in real-time brain imaging, which are supposedly revealing brain processes at the molecular level of neurochemical interaction, maybe these detractors of philosophy and art have a point? As one by one the functions of the black box are purportedly opened up to experimentation, the question posed is, can the neurosciences deliver what philosophy, and art theory, have failed to do, that is, bring about the ultimate materialist understanding of how concepts and sensations are made in the brain? Well, despite the extraordinary speculation surrounding the claims of neurophilosophy and neuroaesthetics, the answer to this question, for the most part anyway, seems to be far from affirmative at this stage. On the other hand, it is important not to underestimate, even in this ostensibly post-Foucauldian era, the continuity of the blind rationality of royal science. It is important not to confuse a Deleuzian neurophilosophy with the imposition of the kind of rational essence of thought (and sensation) we see emerging in the many offshoots of cognitive neuroscience. There is still much potential for rhizomatic mixtures, but as Rosi Braidotti points out, although philosophy, science, and art are all caught up in the ontology of becoming, science is more on the side of actualization than it is the virtual.[16] It is therefore important not to confuse Deleuze and Guattari's chaos brain with a royal brain that is a rational, conscious, moral, cognitive, and all-knowing subject, that is, the dreaded humanist brain. Deleuzian neurophilosophy is a much "subtler tool" for "probing the intellect."[17] It is more attuned to the virtual.

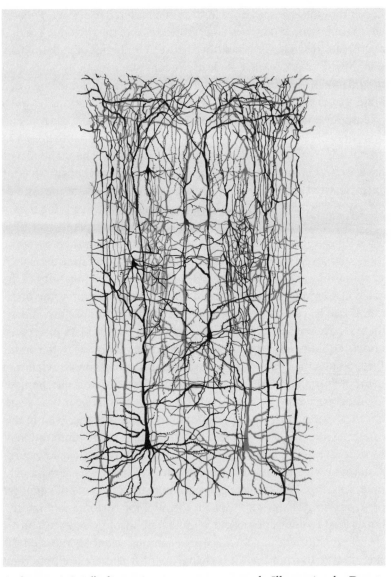

A rhizome? Cajal's discontinuous neuron network. Illustration by Dorota Piekorz.

Saying that, science, and the neurosciences in particular, can also be regarded as working in the virtual in the sense that the propositions forwarded about assumed conscious cognitive processes, such as perception, memory, belief, attention, and opinion, are often problems hypothetically posed (or imposed) on observations of complex neural activity. This is nonetheless a very different rendering of the virtual to the one Deleuze and Guattari were looking for in a brain where the vital ideas of philosophy, art, and even science were not conformed to mental objects or the formation and communication of opinion. The chaos brain of *What Is Philosophy?* is not constituted as an object of science, as such, but rather in a continuous production of thought being made in the "deepest of synaptic fissures, in the hiatuses, intervals, and meantimes of a nonobjectifiable brain."[18] This is arguably another sense of the nomadic space we first encountered with the rhizome brain—a space where thought can escape neural correlations with homogeneous mental states and become heterogeneous.[19] It is also imperative not to mistake the rhizome or chaos brain for alternatives to the rational brain. Neither celebrates irrationality nor exists in all-out chaos. Deleuzian chaos is nothing like this; rather, it is composed in the infinity of possibility. Both brains (rhizome and chaos) are, in the context I wish to explore them in this chapter, interferences that occur in the gaps between the cerebral inventions of philosophy, science, and art. To this end, the often criticized discreteness of the three planes cast by philosophy, science, and art, which feature throughout *What Is Philosophy?*, comes in use insofar as the parting of each plane not only provides a sense of where the dividing lines are but also offers insight into where interferences might potentially occur—*in between the gaps.*

TWO BRAINS, MANY RHIZOMES

It is significant to grasp the continuities, as well as the discontinuities, between the rhizome and the chaos brain. Before revisiting the latter, let's begin by exploring the much-celebrated, but perhaps often hyperbolized, brain made of grass—*not trees*—a decentered brain that resists all kinds of microfascism. It is the rhizome brain that arguably brings Deleuze and Guattari to the attention of a mostly enthusiastic audience of digital theorists in the 1990s. It has certainly inspired philosophers, artists, and political activists alike to experiment (with varying success) in mixtures, smooth spaces, and decentered resistance in the age of networks (digital and neural). In terms of nomadic thought, the brain made of grass may well have been borrowed from early anatomies of the synapse, but unlike the cognitive neurosciences, which have endeavored to replace the transcendent gods and Cartesian ghosts that haunted the brain with a material synaptic gap full of representational mental stuff, the rhizome importantly establishes a relation to chaos: to the event. The rhizome crucially never becomes a function of thought as such. It replaces Cajal's butterflies of the soul (located in the neuronal structure) with the sound of a thousand flapping butterfly wings, none of which have a precise location.

It is, like this, that we find that the rhizome brain has more similarities with the chaos of the ocean brain than it does with the computer brain. It is a nonrepresentational brain: a brain that does not think in images or even traces of an image. To be sure, representational thinking is just another word for *state philosophy*—the *arborescent model* of thought that became the enemy in *A Thousand Plateaus*. This model has a very long history in theories that fall back on the notion that the brain is a storage and retrieval area for representations, impressions, encoded descriptions, or brain traces, dating back to Plato's wax tablet metaphor.[20] Although the notion of a memory storehouse, full of perceptions and opinions, has grown in sophistication, with some cognitive neuroscientists now assuming that mnemonic experiences, for example, are stored as changes in strength at many synapses within a large ensemble of interconnected neurons,[21] tree logic still underpins these cybernetic brain systems—a tree logic that, Deleuze and Guattari contend, is all about "tracing and reproduction."[22] Indeed, the idea that the brain stores mental images, sounds, or smells as representational traces, or repeatable patterns of neuronal firing, does not explain how remembering the way home, for example, requires no such retrieval of a representational mental image or utterance in the mind, just a capacity not to forget to act or react.[23]

What if you were suddenly to forget the way home? Is this a problem of memory storage? Not necessarily; as Michael Schillmeier puts it, remembering is a refeeling, or remaking, of social presence.[24] It is not a cognitive process stored someplace but a noncognitive, Proustian *re-member-ing.*[25] Re-membering is not a reproducible or representable phenomenon. The thing remembered is not a ready-made "lurking in the dark recesses of memory and language."[26] It is not a trace—"tracings are like the leaves of a tree"[27]—or an overcoding of neuronal patterns. Though Deleuze and Guattari themselves fall back on cognitive distinctions made between short- and long-term memory (preferring the seemingly rhizomatic actions of short-term memory to the imprints, engrams, tracings, or photographs of long-term memory),[28] they do not focus on memory storage or the retrieval of representations. They draw attention instead to the lack of contiguity and the important role of forgetting in re-membering. Even in the case of long-term memory, whatever it is that is remembered continues to act in such a way as to spring to mind or disappear *without a trace*. Indeed, the rhizome brain is altogether different from the tree brain. It is not a whole system; it is an arrangement, an assemblage.

At first glance, the chaos brain does appear to be a very different proposition to the rhizome. The promises of a brain capable of producing a nomadic science seem to have all but dissipated into the distinct planes Deleuze and Guattari introduce in *What Is Philosophy?* The chaos brain is still confronted by the chaos of the event, but the cerebral enunciations of scientists, philosophers, and artists take on a discreteness that seems to, initially, undo the philosophy of mixture we encountered in the schizoid analysis of both *Anti-Oedipus* and *A Thousand Plateaus*. Nonetheless, like the rhizome, the chaos brain of the philosopher, scientist, and artist does not become a unity but rather a junction point between itself and chaos. It is almost, in this sense, the prequel to schizoid analysis, many of the nomadic qualities of the rhizome eventually emerging in the closing chapter on the indiscernible expressions of the chaos brain. Indeed, a series of rhizomatic lines of flight to follow in *What Is Philosophy?* need to be noted:

1.   There is, despite the discreteness of the artist's sensations and the philosopher's concepts, a relation established between feelings and thoughts in the chaos brain. It is, as such, a brain that is no less sensation than it is concept. In other words, it is a brain that feels sensations as much as it thinks concepts.

2.   Like the rhizome, the chaos brain refuses to have mental models imposed on it. Ideas do not become the mental objects of an

objectified brain. Philosophy, art, and science instead become the three aspects under which the *brain becomes subject.*

3. The chaos brain is not the brain of the phenomenologist; that is, it is not "the Man who thinks beyond the brain." The thinking person is the ascent of phenomenology beyond the brain toward a *Being in the world.* However, Being in the world does not help us to escape fixed ideas, and opinions, which, as we will see, are the adversary of nomadic thought. Here we see how Deleuze and Guattari's incorporeal materialism makes an appearance. *There is no brain behind the brain.* But then again, there is that problematic statement: "*it is the brain that thinks, not the mind or the person.*" Such a concept of the material brain prompts a series of interrelated problems.

  A. The question of what constitutes *subjectivity in the making* is answered by direct reference to the brain rather than the human. Indeed, as Andrew Murphie points out, it would appear that the brain proves to be the key substitute for the human in Deleuze and Guattari's ontology.[29]

  B. Is it really the case that the brain is the mind itself? The brain may well say "I," but the "I" is an Other etched onto the immanent brain. The "I" of the brain is not transcendent. By ridding the brain of the phenomenological "I" (mind and person), we must be careful not to reintroduce a transcendental brain through the back door. A brain is just a surface that is incorporeally coated.

4. The chaos brain confronts chaos before it casts out three distinct planes of enunciation. So, significantly, it is these three planes, not the chaos brain itself, that seem to make the cerebral expressions of artists, philosophers, and scientists discrete from each other. Brains are not centers. They are merely points of exchange in assemblages of sense making.

5. Finally, the chaos brain is open to three kinds of interference. Indeed, eventually *What Is Philosophy?* points toward a third interference featuring a chaos people to come that could have walked straight out of a chapter of *A Thousand Plateaus.*

CHAOS AND THE NEUROIMAGE

Central to the neuroscientific claim to have the answers to the big philosophical questions in its grasp is the neuroimage. It is here, in the glowing heat maps of the fMRI scan, for example, that some neuroscientists claim to be able to observe human culture. The interferences between concepts and sensations have not, however, been that easy to discern. The categorizing of mental objects has proved to be little more than an unstable series of reference points thrust into chaos. The flashes of color that light up the fMRI scan provide unpredictable coordinates (and variable relations) linked to more or less metaphysical assumptions of what constitute the psychological categories of memory, attention, perception, and emotion. The neuroimage has not proved to be the wholesale capture of these mental objects. Indeed, in many ways, neuroimaging has dragged the mental models of the cognitive sciences kicking and screaming back into the chaos that haunts the brain. The cognitive proposition of working memory, for example, is not so much revealed by fMRI as it is caught floating above the brain image like a Cartesian ghost. Cognitive science is haunted by the chaos exposed by the neuroimage. The search for single modules and localized brain regions controlling such things as working memory has led to multiple brain components, activated and spread over many brain regions, taking part in numerous cognitive and noncognitive tasks in a variety of contexts. Although claims concerning objective and localized mental states are widespread in neuroculture, these states arguably amount to little more than what the biophysicist Robert G. Shulman calls "phrenological fMRI."[30] Once the black box of the computer brain has been opened, it would seem that a further labyrinth of black boxes begins to appear. The neuroimage does not show us a brain that is localized, compartmentalized, or computerized. It is instead a biophysical brain that responds to, and supports, a body's desire.

The cognitive neuroscientist may well have captured pieces of chaos in the neuroimage and turned them into bits of *Nature* that can be intervened into. Brain-imaging technology has undoubtedly helped the neurosciences go beyond anatomy, making the movement of brain function visible and allowing for interventions to be observed as they happen in brain chemistry. Significantly, perhaps, sensations have been traced from the lab rat's whiskers all the way to specific locations in the brain (see chapter 5). The neuroimage thus shows us something of *what can be done to a brain* via exposure to sensory environments. But any hope of finding the essence of the computer brain in the neuroimage, or, for that matter, its production and storage of concepts, has yet to materialize in neuroimaging research. The engineering metaphors of cognitive neuroscience were perhaps always destined to end up swimming about in an ocean. Indeed, the ocean brain is much nearer to the chaos brain, because its openness to endless possibilities and infinite chaos coincides with the brain as an event. The question of *what a brain can do* is therefore best posed by way of concepts and sensations that do not subtract from the infinity of the event.

APPROACHING INCORPOREAL SURFACES

In *What Is Philosophy?*, the first thing to deterritorialize is not the brain but the person. The chaos brain is certainly not a person, or a mind, but a material surface. The subject in the making becomes a brain that is not transcended by anything other than its own surface, as such. The idea of the surface is nothing new. Following Ruyer, Deleuze and Guattari find nothing beyond or behind the surface of the brain. Contrary to locationist and phenomenologist tendencies in neurophilosophy, there is no mind map of the brain or metaphysical presence hovering over matter other than that which is found on the tissue surface. It is indeed possible to grasp why Deleuzian incorporeal materialism would wish to jettison the concept of a transcendent and phenomenological brain. This is because phenomenology "goes in search of original opinions which bind us to the world."[31] It looks for them in the beautiful and the good experienced in the perceptions and affections that would awaken us to the world. Instead, Deleuze and Guattari look to use concepts and sensations to respectively route around the opinions and perceptions that produce the inner "I" in the world. Like this, percepts replace the limits of perception, whereas affect is extracted from affection.

An important question that needs to be addressed, it would seem, is how to distinguish between the incorporeal surface of the brain and the materiality of approaches in the neurosciences that devolve all mind processes to the synapse. Indeed, are we not, by looking at the brain only as a surface, at risk of agreeing to a reductive philosophy of the neuron? Reduction is of course one way in which it is assumed that philosophical and scientific approaches to sense making might become unified, but this is certainly not my intention here. Interferences must not become unifying forces. To begin with, by saying that it is not the person or the mind but the brain that thinks, we must avoid concurring with aspects of cognitive neuroscience, which, while claiming to use a bottom-up empirical approach, increasingly aided by neuroimaging technology, can never really escape the top-down application of mental objects imposed on neural activity. In other words, by proposing that he has observed how memory works at the level of the neuron, or neuron network, the cognitive neuroscientist has also supported a ready-made concept of how the brain remembers. A Deleuzian relational ontology would surely never accept such a cognitive position pinned together by off-the-peg psychological categories, particularly those models originating in engineering metaphors that limit analysis to the relation an inside model of the mind has to an outside world.

There is a need to approach the interferences between science, philosophy, and art with care, that is to say, wary of a further tendency to ignore the potential for surface-to-surface, or intercorporeal, relations established between human and nonhuman worlds in favor of aborescent evolutionary divisions. To be sure, the chaos brain is a resolutely materialist philosophical invention, but it must not exclude the exteriority of lived encounters with events. It is prudent, as such, not to make the error of reducing a broader understanding of *social* relations, such as those that might be realized in the empathic sharing of feelings, to the interiority of neuronal activity assumedly determined by genetic mechanisms. By grasping mirror or empathy neurons, for example, as the shapers of the social and cultural milieu of the human, we run the risk of encountering a brand of sociobiological neuroscience that, as Tiziana Terranova points out, imposes the selfish gene on neuronal interactions. Although we should not altogether ignore the potential of brain empathy as a component of a radical social and cultural theory, the focus on mirror neurons

> tends to rely on sociobiological theories [which provide a] recoding of networked subjectivity onto the figure of the manipulative primate, whose social intelligence is imitative in nature and where

imitation is basically the key to social manipulation by a self-interested, calculative subject endowed with strategic rationality.[32]

Indeed, the supposed recursive actions of mirror neurons (or empathy genes) have been cited by Ramachandran, among others, as one way in which human societies and cultures are supposed to have evolved through competition with other animals.[33] The duplicitous and ironic nature of this human-centered predisposition in the neurosciences perhaps comes into full view in Arthur Kroker's observation of the "objectification of animals by the scientific gaze."[34] In short, what kind of empathy is it that requires vivisection as a condition for an epistemological grasping of the human capacity to share feelings?[35] If we are to produce ethical interferences with this kind of neuroscientific version of empathy, which unfeelingly partitions human and nonhuman worlds, then there is a need to redefine what it means to share feelings. Herein, intersubjectivities need to be grasped as traversing species lines and becoming empathically related to each other in an inclusive environment. This is a vitalist interrelation wherein the sharing of feeling expresses the capacity to affect and be affected by an environment.[36]

OPEN TO THREE KINDS OF INTERFERENCE

It is, nevertheless, despite the limitations of neuroimaging, important to remain open to the propositions of the neurosciences—to never be afraid to look for, and experiment with, interferences that crop up in between the functions of science, philosophical concepts, and artistic sensations. This is consistent with Deleuze and Guattari's final analysis that a nonscientific reading of science is an entirely valid course of action for philosophers. Therefore, a number of methodological moves can be drawn on to tackle the problems related to borrowing from the neurosciences so that interferences between philosophy, science, and art can be affirmatively prompted into action without becoming beholden to the reductive forces of the royal sciences. First, we can assume Manuel Delanda's intensive science position; that is to say, we accept that there is no such thing as *science in general* and make instead a distinction between scientific propositions that are either taken from the hypothetical cutting edge of science or those that we might take from hard science.[37] In the case of the former, a proposition like the mirror neuron, for example, could indeed prove to be an interesting, yet hypothetical, notion of how empathy is transferred between social brains. In the case of the latter, the existence of neurons might come under revision but is generally unlikely to be proved wrong in the future. Either way, science only serves as an illustration of an ontological viewpoint grasped in the historical movement between virtual and actual propositions. In this light, mirror neurons can perhaps be poached by philosophers and used as a convenient but conjectural illustration of how empathy contagion might be processed at just one level of a complex layering of interaction, including biological, social, cultural, and political layers. Indeed, the fairly recent identification of neuronlike cells in the gut might similarly suggest the existence of a second brain.[38] The point is that these kinds of interferences are grounded in ontological commitments to relationality, not temporary epistemological ignorance.

A second approach (the one I develop here) would be to view philosophy, science, and art as three discrete enunciations of sense making, which can, as already set out, overlap and interfere with each other in productive ways that do not constitute a unity. While philosophical concepts, scientific functions, and the sensations produced by artists might be regarded as somewhat incompatible, there is the potential for creative inventions to emerge from three kinds of interference (*extrinsic, intrinsic,* and *nonlocalized*). Like this, it will be important to grasp the differences between concepts, functions, and sensations and look for new lines of flight that might occur when they overlap each other. The interference therefore "offers insights into the future of philosophy . . . art and science, and ultimately of thought itself as a confrontation with chaos,"[39] but it is also grasped here as a political act of piracy. It is not intended to be faithful to the original. After all, what work of science, art, or philosophy is ever truly original, that is, not a product of an imitative encounter with another example, which is, in turn, an imitation, and so on? A philosopher can, however, freely interpret a work of art or a function of science in such a way as to create a new concept from the original without becoming locked into discursive formations and existing referents. Indeed, although the interference is always imitatively related, it can produce a new referent from what is poached, albeit limited, in the case of intrinsic and extrinsic interferences, to never becoming a scientific function of knowledge or sensation of art.

That is to say, what is poached can only ever become a concept of a function or a concept of a sensation.[40] To produce this kind of novelty, interferences need to oscillate in all directions. The philosopher steals from the scientist as she does from the artist, and likewise, the scientist is free to steal from the philosopher and the artist. To be sure, neurophilosophy and neuroaesthetics are examples of piracy on behalf of a scientist who pays very little respect to an original. The interference is not, as such, assumed to be interdisciplinary. The point is to avoid unity through rapprochement. The interference is not supposed to be a harmonious relationship; it can, and should, be complementary and antagonistic in equal measure.

The eventual aim is to use the interference as a conceptual tool that routes around fixed ideas, opinions, and regimes of perception. For example, in the first part of this book, the science of computer work, neuropharmaceutical intervention and the early development of brain wave technology are forced into interferences with the sensations of the dystopian novel. This leads to a concept of power in which populations are controlled by way of noninvasive interventions into brain chemistries. This is a concept of power that puts the neuron doctrine to work as part of new managerial efficiency drives focused on the entrainment of affective rhythmic brain frequencies associated with work, attention, and consumption. In the second part, interferences between the neurosciences and philosophy are used to think through a concept of brain freedom, that is to say, an assemblage brain unencumbered by phenomenological and locationist tendencies in neuroculture. As Arthur Kroker argues, the "human brain has always been deeply relational and profoundly ecological."[41] We need to consider the various posthuman exits that become available to us in a neurodiversity found beyond a model of the brain trapped in the cranium or computer metaphor. At very least, we need to consider "the mysterious, enigmatic connection between brain matter and the full human [and nonhuman] sensorium— vision brains, tactile brains, gut brains, ear brains."[42] In the second part of the book, I similarly evoke Tarde's pansocial monadology as a way to grasp sense making liberated from this atomistic prison cell to reveal societies of cells in which it is not the cell itself but the wider maelstrom of cellular interaction with matter that coevolves in a sensory environment. This combination of Deleuzian and Tardean ontology produces a relational assemblage theory of the brain, so that *the brain that thinks* is not merely a function of the neuron, or indeed rendered to localized brain regions of interacting cells, but instead becomes an encounter with the event—the infinite chaos. This is, in Tardean terms, a monadology of microbrains, not reducible to the cranium, the brain, or the neuron but to societies of organic cells and inorganic matter encountering each other in assemblages of sense making. This means, ultimately, going beyond the extrinsic interference, which produces a "resonance" between planes, and intrinsic interference, which "takes philosophy outside philosophy, say, into science or art or makes it difficult, even impossible, to decide to which field a given concept belongs,"[43] toward an interference that says *no* to itself, that is to say, "nonlocalizable interferences [that] take us beyond any given field—art, science or philosophy—even from within." This is an interference that becomes "quantum-like."[44]

Before we can even begin to contemplate the nonlocalizable interference, we need to better establish the ontological considerations Deleuze and Guattari give to the brain's encounter with chaos.

THE WORK OF CONCEPTS, FUNCTIONS, AND SENSATIONS

It is the work of artists, scientists, and philosophers to "tear open the firmament"—the umbrella that shields us from chaos—and "plunge" into the infinity that confronts us all.[45] Chaos, in this sense, is not a disorder or nothingness; *chaos is the virtual*. It is the void of endless possibilities the chaos brain plunges into, without consequence, consistency, or reference.[46] The differences between science, philosophy, and art are not simply about conflict between each discipline but rather about the discreteness of the enunciations that follow the confrontation with chaos, which add consequences, consistency, and reference to the world. Chaos is indeed what situates the assemblage brain, making it a junction point between chaos and matter. It is the work of artists, scientists, and philosophers to therefore use this junction point to cross over into the chaos and meet it head on—to return with an articulation of the events experienced.

The philosopher's creation of concepts "proceeds with a plane of immanence or consistency."[47] Philosophy endeavors, as such, to retain the infinite speed of chaos and give consistency to the virtual. The philosopher's concepts can be looked at in several ways. First, they are relational. Concepts have components that relate and overlap. The concept is not a transcendent idea; it is an assemblage. Second, concepts become inseparable and heterogeneous. Yet, paradoxically, they become consistent. In this paradoxical state, the components of a concept can become indiscernible. "There is an area ab that belongs to both a and b."[48] Third, concepts have intensive features. They are points of coincidence, condensation, and accumulation. Unlike the spatiotemporal coordinates of science, the concept is an intensive ordinate. For example, an ordinate relation of the bird concept is not about genus or species line but about bird postures, colors, and song. The concept of birdsong, for example, is a variation—a refrain set to become a multiplicity, a line of flight. The birdsong concept sings the event, not the essence or the thing. This is, as we will see in the introduction to the second part of this book, the birdsong as event, a social relation—not a memetic bird, that is to say, the culture of birds guided by a genetic-like mechanism, but a Tardean bird.[49] Finally, in philosophy, the concept is not intended to form an opinion but to surpass it.

In science, knowledge is not a form. It is a function; or rather, knowledge is *made* functional. It is put to work as a coordinate, a correlation between variables or, more specifically, in cognitive neuroscience, a neural correlate between brain activity and mental states. To think through how these functions of knowledge develop, Peter Gaffney usefully refers to a *gap* that is not unlike Delanda's intensive science. Indeed, science, just like philosophy, finds itself "in the gap" between the virtual and the becoming-actual (or mature of the concrete).[50] Both are thus situated within an indeterminate *zone of immanence*:

> The capacity to interrogate matter, to render signs from experimental data, is only possible by virtue of this smallest distance between a continuously renewed problem—the force of the virtual, as such—and the imminent solution [the actual].[51]

The problem can, as such, always escape the solution. In other words, in the process of science becoming mature, the problem will tend to bifurcate and follow new lines of flight. So, can we ever say that science ever truly becomes mature?[52] Significantly, the answer provided in *What Is Philosophy?* is that it does not come into being or find absolute unity. The new paradigm becomes no less contested than it was when immature.

Unlike the intensities of philosophy, science is an extensive system, with references made to states of affairs and bodies. As already mentioned, neuroscience has a constant dream of capturing a bit of chaos in the neuroimage so that it becomes Nature. Like this, the neuroscientist tries to make evident the chaos into which the brain itself (as subject of knowledge) plunges. In contrast, philosophy exists outside of coordinates or correlates, in virtual neighborhoods and zones. Concepts are indeed junctions, movable bridges, or detours. Whereas science produces a series of propositions, a discursive formation, *it is the goal of virtual philosophy to refuse to become a discursive formation*. Its goal is instead to expose the nondiscursive—the event and affects.

Other distinctions that need to be made at this point are established between the relation philosophic concepts, artistic sensations, and scientific functives have to events. Here the difference between the variations of concepts and the variable application of functions to a state of affairs becomes even more apparent. Concepts have events for consistency, whereas functives (the components of functions) have states of affairs or mixtures for reference. So that, on one hand, in the creation of concepts, philosophy extracts a consistent event from the state of affairs. On the other, science works to "actualize . . . the event in a state of affairs, thing, or body that can be referred to."[53] To reinforce this point, Deleuze and Guattari refer back to the Stoic distinction made between states of affairs, in which events are actualized, and mixtures of bodies wherein incorporeal events rise like vapor from the state of affairs.[54] The difference here is found in another distinction we can make between the scientific relation established between variables (used to set a limit in the state of affairs intended to slow down the mixture of bodies) and the inseparable variations of the plane of immanence (virtuality and mixtures) that philosophy intervenes in.

The artist's sensations also need to be differentiated from concepts and functions in terms of their relation to events. Take, for example, the sensations of sculptures, paintings, novels, or music and note how they never actualize the event but instead incorporate or embody it. The sensation is in fact neither virtual nor actual. It is always the possible! The sensation of the sculpture gives a body to the event, providing it with a universe of possibilities in which to live. Sensations are what the artist composes with her materials, offering contemplations and enjoyments,[55] but they also produce a universe that constructs limits, distances, proximities, and constellations.[56] An artwork is a bloc of sensation that routes around perception and resemblance by putting affects and percepts into motion.

It should not go unmentioned at this point in the discussion that the interferences produced when the artist's sensations and philosophical concepts intersect are somewhat perplexingly dealt with in *What Is Philosophy?* Conceptual art is explicitly (and somewhat surprisingly) rejected because, as Deleuze and Guattari contend, when art tries to become *informative,* it also becomes unclear as to whether it is a sensation or a concept.[57] Indeed, it is not in the artwork itself but in the "opinion" of the spectator that the sensation is, or is not, manifested. This is not the machinic distribution of affect mediated through the experience of sensation but a "subjective act of signification."[58]

Perhaps it is the case that the artist who endeavors to mix signification with sensation risks losing some of the affective critical clout of the artwork, but surely the inclusiveness of the spectator in the interferences produced when concepts and sensations collide is something that needs to be encouraged. I return to this point later.

GIANTS, PERSONAE, OBSERVERS, AND FIGURES

Evidently, concepts, functions, and sensations are enunciated through the names of certain *giants* who themselves stand on the shoulders of other giants, and so on. These are examples taking part in processes of imitation and counterimitation. To be sure, scientists, philosophers, and artists seem compelled to acknowledge or refute the giants who come before them. To justify their own standing, perhaps, they make further use of these names to pin down the lines of flight emitted from their chaotic adventures. Philosophy uses these names to mark out bifurcating events so that, for example, Kant breaks with Descartes.[59] Science similarly uses the names of these giants to cut off a paradigmatic pathway so that, for example, Newtonian mechanics is succeeded by Einstein and subsequent subatomic explorations into quantum mechanics. The paradigmatic path of the neurosciences is no exception. In the case of the development of the neuron doctrine, there were those who followed the reticularist view of the brain, forwarded by the

Italian physician Camillo Golgi. The reticularists became cut off from the prevailing neuronists, led by the Spanish pathologist Cajal, who began to lay down a new pathway eventually leading to the neuron doctrine. The discontinuous neuron ultimately becomes actualized in hard science. However, there is always a possibility that the reticularists might one day return to haunt the neurosciences. Indeed, Golgi may already be back in the gap.[60]

Art, of course, has its own giants, neatly attributed to various movements that cut a fashionable path through a synthetic history. Signposts are dotted along this path, some with faded or forgotten names, others revived and brought back to life from the dead. As already mentioned, the most controversial of these art pathways is perhaps followed by Deleuze and Guattari to a junction whereby the emergence of abstract art and conceptual art seems to promise to bring art and philosophy together.[61] The shadow of Duchamp's giant cannot be ignored here, because he is seen by many to have brought together two paths on which sensations and concepts once traveled alone. But again, *What Is Philosophy?* appears to suggest that concepts and sensations do not mix well. They are certainly not regarded as interchangeable substitutes. On one hand, the abstract artist refines the sensation. This is the dematerialization or spiritualization of matter in which the sensation of Turner's chaotic seascapes, for example, becomes an early sensation of the concept of the sea. Conceptual art, on the other hand, dematerializes through generalization. Here once more Deleuze and Guattari seem to cautiously question whether conceptual art leads to either sensation or concept. Indeed, they explicitly reject conceptual art on the basis that it is more the latter than the former. The role of the audience in practices of signification becomes even more prominent in conceptual art, they contend. Given the information the audience receives from the artwork, it is they who determine if the concept will be materialized into a sensation. That is to say, it is *they* who decide "whether it is art or not."[62] The problem here for Deleuze and Guattari is that conceptual art thus produces signification, not sensation. This is perhaps a surprising denunciation given that conceptual art is now the condition for contemporary art, and it presents a startling contradiction with Deleuze and Guattari's role as the philosopher kings of contemporary art. This is a point of contradiction nicely captured by Stephen Zepke:

1. Deleuze and Guattari explicitly reject Conceptual Art.
2. Conceptual Art is the condition of Contemporary Art.
3. Deleuze and Guattari are touted far and wide as the philosophers of Contemporary Art.
4. ... Huh?[63]

Perhaps contemporary art has been decidedly selective with regard to its choice of Deleuzian brains, preferring rhizomes and nomads over the discrete divisions established between concepts and sensations in *What Is Philosophy?* Yet Deleuze and Guattari contend that art as sensation retains a political power that art as signification cannot. The former is imbued with autonomy, whereas the latter, they feared, would collapse into immaterial capitalism. So further consideration undoubtedly needs to be given to the interferences produced when art comes into contact with market capitalism. From the comfort of our current position, we can perhaps see more clearly where the interferences between art and market are. Looking back at conceptual art's development since the early 1900s, it is indeed evident that it has become increasingly part of an art system made up of art critics, uber-dealers, advisors, bankers, and now superrich oligarchs. It is they, not the audience, who decide if a concept is indeed art (or not) by attributing a discourse and a price tag to it. The pathways cut by the giants of art cannot, it would seem, be readily detached from the cultural circuits of capitalism in which this system thrives. This is a system that has long since expanded beyond the artist's studio and the gallery space. As Deleuze argued, "art has left the spaces of enclosure in order to enter into the open circuits of the bank."[64] Fairly recent pictures of oligarch superyachts moored outside the Venice Biennale in 2011 are just further testament to the latest embrace between the art system and the rich. Indeed, the post-2008 crash emergence of a new superrich has produced an art system that even Charles Saatchi calls "too toe-curling for comfort."[65]

The giants of conceptual art have become fully intertwined with capitalism; but now, it seems, artists are looking for an escape in postconceptual art, which attempts to create new tools to resist the control society, as, for example, Ricardo Basbaum's work claims to do.[66] The postconceptual artist needs to steal back the concept from the art system, he argues, and control the right to frame his own ideas outside of market control. Beyond this, perhaps, as Zepke argues, conceptual art needs to be followed by a nonconceptual art. But how safe is the autonomy of the sensation? Indeed, in times of neurocapitalism, and in the practices of neuroaesthetics in particular, the political autonomy of the sensation will come into increasingly antagonistic relation to the functions of a scientific knowledge working for the market. The political harnessing of the affective experiences of the artist's sensation is of course nothing new, but it is becoming more scientific. To defend against any neuroscientific attempt to make the location of aesthetics in the brain (or elsewhere) visible in a hard science, artists must continue to mobilize affects that disrupt control. The ultimate interference that art might contribute toward may indeed be the invention of a non-art that altogether avoids the locationist tendencies of neuroculture.

The problematic discreteness between concepts, functions, and sensations may also be explained by the limitations imposed on the potential of interferences by the giants themselves. Like this, giants are constrained to *extrinsic interferences* that can produce only a resonance between planes:

> A first type of interference appears when a philosopher attempts to create the concept of a sensation or a function (for example, a concept peculiar to Riemannian space or to an irrational number); or when a scientist tries to create functions of sensations, like Fechner or in theories of color or sound, and even functions of concepts, as Lautman demonstrates for mathematics insofar as the latter actualizes virtual concepts; and when an artist creates pure sensations of concepts or functions, as we see in the varieties of abstract art or in Klee. In all these cases the rule is that the interfering discipline must proceed with its own methods.[67]

In addition to the contested role these giants play in articulating links between systems, marking out paradigmatic paths, and the breaks and knots in syntax, there is also the widespread unleashing of subjectless forces into the chaos. Indeed, to some extent, the problem of the discreteness of each plane may be overcome by the forces of a far more subtle *intrinsic interference* that produces sliding from one plane to the next. It is perhaps more prudent to therefore follow these *virtual sense makers* than it is to pursue giants. Here Deleuze and Guattari locate the *demons* that "swarm throughout sciences," returning the results of interacting variables, counting and judging speeds, and observing the limits of technology and data capture. In this way, Laplace and Maxwell's demon boldly goes where no subjectivity has gone before—certainly beyond the observable field of human phenomenology. But demons have their limits, too. They are not to be misconstrued as the *All-Seeing Eye*. They are instead like an eye at the summit of a cone.[68] They have many blind spots. These demons therefore become the *partial observer* in relation to functions within systems of reference. Scientific propositions always require extrinsic demons that do not act; they merely perceive and experience the chaos.

Philosophy also has its own demonic forces. The philosopher brings to life various personae through which concepts can live, as such. However, the conceptual persona *lives* only in relation to the fragmentary concepts located on a plane of immanence. They should not be mistaken for a person. They are *larval subjects*. Through the lives of the conceptual persona, concepts are thought, perceived, and felt. Indeed, the production of concepts requires these intrinsic conceptual personae to be able to make interventions. Plato used Socrates for such purposes, and Nietzsche introduced many personae: Zarathustra, Dionysus, Overman et al. In *What Is Philosophy?*, the conceptual persona of the idiot is well deployed.[69] The idiot is contrasted with the "Schoolman" or public teacher in the sense that he is never resigned to the *facts* or to historical truths.[70] Indeed, it is the questions asked by the idiot that unleash the creativity of the absurd, the lost, the discarded, the anomalous, and the dysfunctional. However idiotic these questions might appear to be, they create an alternative to the taught concepts of the public teacher, challenging their rationale by slowing the concept down and making it stutter. These anomalous personae produce important interferences between philosophy and neuroscience not least because they help to reveal the normalizing tendencies of neurocapitalism (see chapter 3). We might say that the inattentive subject (the ADHD child), like the forgetting body of Schillmeier's demented person, is a conceptual persona that puts the "powerlessness of a political voice [back] into politics."[71] That is to say, he can serve to make us rethink our position and ways of doing things outside of the neurotypical range.

Although at the extrinsic level of the interference, concepts and sensations are kept at a resonating distance, preventing their full mixture, there is a closer relationship, it would seem, established between the *subjectless* forces of art and philosophy at the intrinsic level. Like this, the aesthetic figures of art playfully interweave with the philosopher's conceptual personae. Just as the latter routes around opinion and fixed ideas, the aesthetic figure of the novel, painting, sculpture, and music produces percepts intended to surpass ordinary perceptions and affects to be extracted from affections. Nonetheless, aesthetic figures are not the same as conceptual personae. Again, we need to ask where the philosophy of mixture of *A Thousand Plateaus* has disappeared to. There are a number of differences to note. To begin with,

> it may be that they [conceptual personae and aesthetic figures] pass into one another, in either direction, like Igitur and Zarathustra, but this is insofar as there are [only] sensations of concepts and concepts of sensations.[72]

Moreover, whereas conceptual personae take on a Promethean role, bringing concepts to life, aesthetic figures are more like vampires, looking for a body in which to live. They can only live through their percepts and affects. So, art and philosophy traverse chaos and confront it, but on very different planes. While "art thinks no less than philosophy," it *thinks* as a bloc of sensation.[73]

Then there is a further difference between sensorial and conceptual becoming. The latter is heterogeneity grasped in an absolute form (heterogeneous consistency), whereas the former is an otherness caught in a matter of expression.[74] Like science and philosophy, art therefore finds itself in the virtual–actual gap, but the artwork does not endeavor to actualize the virtual event. Instead, it incorporates or embodies it: it gives it a body, a life, a universe of possibilities. It provides the virtual with a figure. But let's be clear: these figures have nothing to do with resemblance—they are the possible conditions of bringing composition to life beyond representation.

Last, despite a commonness established between the brains of scientists, philosophers, and artists, there are distinct differences in how their experimentations sense chaos and how their thoughts and sensations become an expression of it. To be sure, the commonness of the chaos brain comes about because conceptual personae, partial observers, and aesthetic figures all partake in this enunciation of chaos. Science, philosophy, and art are indeed similarly articulated in the sense that they are experimentations in thought. *There is certainly no creation without experimentation,* and there is as much creation in science and philosophy as there is in art. All three are equidistant, as such, to chaos, and despite their creativity, they are more often than not overwhelmed by it.[75] The difference here is again found in the way that science expresses the reference to a state of affairs, philosophy articulates the consistency of a concept, and art composes the sensation. To this extent, Maxwell's demon, Nietzsche's Zarathustra, and Mallarmé's Igitur become *sensibilia* who plunge into the chaos and return as the voice of enunciation. But amid the observations of partial observers, the ideas of conceptual personae, and the sensation seeking of aesthetic figures, there is an important difference in the way science, philosophy, and art articulate what they have sensed. Zarathustra does not transmit information back from the chaos. He is always already on the chaotic horizon, "circumscrib[ing] a (sympathetic or antipathetic) affect."[76] In contrast, Maxwell's demon adopts a "point of view in things themselves that presupposes a calibration of horizons . . . slowing-downs and accelerations."[77] The demon's perceptions therefore become a quantity of information returned from chaos, while Zarathustra turns to qualities of affect. The artist, in contrast, is always deploying aesthetic figures so as to add new varieties to the world.[78] The aesthetic figure is the medium through which the artist experiments with sensation, drawing the audience into a plane of composition.

The two central questions of this book are littered with *sensibilia* intended to stir up intrinsic interferences between neuroscience, art, and philosophy. For example, in the first part, Aldous Huxley's Brave New Worlders, high on soma, open up an entire dystopian universe of control that explores the extent to which a brain can be acted on. Huxley's dystopian novel gives a body to the dulled sensations of a contemporary docile control society in which the attention deficits of the brain become increasing normalized and quantized to the rhythmic pulses of neurocapitalism. This is a control society determined to a great extent by joyful encounters with marketing. So it goes that everybody is happy to pay attention—to the *right* things. Indeed, in this current year of Our Ford, as Huxley calls it, *everybody's happy now—in everybody else's way.* In the second part, we see how some twenty years later, in his book *The Doors of Perception,* Huxley transforms himself into an aesthetic figure with an altogether different variety of psychostimulant.[79] Like this, he inhabits his own composition, his own aesthetic universe. Psilocybin, mescaline, and, later on, LSD provided Huxley with an escape into an altered brain state *out there* in objective reality, which interferes directly with a neuroscientific plane in which self-identity is always trapped *in there*—in the subjective experience of the individual.

THE THREE PLANES

From *Brave New World* to *The Doors of Perception*, it is the sensation of the drug compound itself that embodies, first, the stratified social classes of the dystopian world and, then, Huxley himself. It is this sensation of the drug that casts the *plane of composition*. Staying with this theme, we can see how, in science, too, drugs are experimented with. It is, like this, that they cast a plane, but the function of the drug does not take on the form of a composition. It becomes instead a coordinate correlating certain causes to various intended or unintended effects or side effects: a kind of drug test. This is the scientist's *plane of reference*, which overlaps with the marketing of big pharma (see chapter 3). Similarly, philosophers do drugs, but they become neither composition nor reference point. They are ordinates: a moment or *plane of consistency* floating on the horizon.[80] There are evidently interferences to be had when the drugs of science, philosophy, and art are taken together. For example, in chapter 5, I look at the work of the neurophilosopher Thomas Metzinger, who considers LSD to be one way by which the illusion of his neuronal self can be altered so that it is well and truly *out there*. Similarly, I follow Huxley's experimentations with hallucinogenic sensations, which led him to a concept of *out there* that Henri Bergson similarly encountered when he, too, routed around the confines of perception.

Before summarizing each plane cast, it is important to add that they are not perceptions or opinions developed as a result of the brain's confrontation with chaos. Although perceptions and opinions seem to prevent thought from dissipating into the void, the chaos brain is in fact the enemy of both. It is significant to note that the three planes are not umbrellas, firmaments, or religions intended to protect the brain from chaos. Tiny slits, tears, and holes are made in the perceptions and opinions that are supposed to shelter the brain from the smooth spaces of chaos. It is only after the chaos brain makes its plunge into the endless possibilities of the virtual that it can return to cast these irreducible planes over chaos. So the brain becomes a common junction point from which these discrete planes emit.

THE CONSISTENCY OF CONCEPTS

Concepts are nothing like the pieces of a jigsaw; "their edges do not match up."[81] Instead of being perfectly aligned, they come together fragmentarily through the throw of a die. Nonetheless, concepts do resonate together as a plane. Once more, chaos does not equate to nothingness. It is the void of endless possibilities; *chaos is the virtual.* Following the confrontation with chaos, the philosopher expresses *variations.* These are infinite differences that have become inseparable from the absolute surfaces and volumes that lay out a secant plane of immanence. Philosophers do not make "associations of distinct ideas, but [create] reconnections through a zone of indistinction in a concept,"[82] for example, zones of indiscernibility between two distinct things, such as human–animal. Moreover, the concept is as much to do about forgetting ideas as it is about forming new ones.

THE COMPOSITION OF SENSATION

The artist brings *varieties* back from chaos, which no longer constitutes a reproduction of the sensory in the organ but sets up a being of the sensory, that is, a being of *sensation* on an "anorganic plane of composition that is able to restore the infinite."[83] Importantly, composition is the work of sensation and not to be mistaken for technique. All artists have a technique. Take an animator, for example, who uses GIF animation (see chapter 4). This is a technique, but the technique needs to disappear into, or become covered up by, the plane of composition.[84] Returning to the drug theme, the composition of the sensation is not to be confused with a chemical substance or material composition either. Indeed, the sensation of soma referred to in chapter 3 must have a *deframing* power that opens up to a plane of composition.[85] Deframing is like a line of flight or musical improvisation that emerges from a refrain. The animator's repetitions, like the dystopian novelist's prose, must sweep up the artwork from out of its technical enclosures, finding an opening and taking up the limitlessness of the plane of composition again. Similarly, the painter, the sculptor—all must make the technique disappear.

So what are the differences in enunciation between the chaos brain of the philosopher and that of the artist? Both are expressions of thinking. But whereas the philosopher thinks through concepts, the artist, Deleuze and Guattari contend, "thinks through affects and percepts."[86]

THE REFERENCING OF FUNCTIONS

The scientist returns from chaos with *variables*. Independent variables are, as follows, intended to slow down the chaos, making dependent variables observable, and measurable, so that the objects of Nature can be slowed down and determined. Specifically, the scientist works by eliminating *variabilities* that are liable to interfere with the variables that are retained. That is to say, dependent variables enter into determinable relations in a *function*: they are no longer links of properties in things but finite coordinates on a plane of reference that go from local probabilities to a global cosmology of possibility.

It is important to update the plane of reference as it applies to the neurosciences today and recent endeavors to make neural correlates between the physiologies of brain function, behavior, and mental states. The move from the actual to the virtual is indeed still haunted by chaos. The sciences that Deleuze and Guattari observed approached chaos by attempting to freeze it, "in order to gain a *reference able to actualize the virtual.*"[87] Cajal's observations of the discontinuities of brain cell structure may indeed be seen as a fantastic slowing down of chaos. However, although his drawings of the infinitely small world of the neuron slowed down a piece of the chaos, it took several decades before the discontinuous physics of the brain could be observed in the synapses. Today, the neuroimage has enabled the noninvasive observation of cerebral energy production and metabolism during sensory stimulation. The image signals produced by these technologies expose the functioning of, for example, the supply of nutrients and blood flow to the brain as well as the electrical activity of neurons. More recently, though, these technologies have been used to correlate brain physiology with a wide range of behaviors and mental activities, resulting in a fragmented plane of reference. Like this, the immature science of brain imaging may, in this new technological environment, appear to mature, but it will never fully mature or, indeed, become a unity, as new limits and lines of flight will emerge from experimentation with mental phenomena that refuse to become localized in the brain.

There is nothing general or unifying about the neurosciences—"science does not carry out any unification of the Reference but produces all kinds of bifurcations on a plane of reference that does not preexist its detours or its layout."[88] To be sure, science cannot actualize the chaotic virtual without appropriating a *potential* from the process of actualization. It is this potential that is distributed throughout the coordinates of a system. As Deleuze and Guattari put it, "the most closed system still has a thread that rises toward the virtual, and down which the spider descends."[89] So scientific paradigms, as Thomas Kuhn famously set out, are where the limits become embedded in the research, to the extent that they appear to become the normal science but will, at some point, become usurped by a new set of limits and potential trajectories, which will eventually become part of a newer paradigm, and so on and so forth. In contrast, philosophy provides a syntagmatic analysis.[90] That is to say, it explores elementary relations on the surface structure rather than the paradigmatic repetitions of the problem. It is important to grasp how it is philosophy that confronts chaos, waiting on its immanent horizon to be carried away, while the intercessors of science are required to slow it down to calibrate its forces. Indeed, chaos problematizes the scientific act of observing by demanding a leap into ontology. Although science brings its own laws and principles into contact with the problem, chaos will always haunt it. It is the chaos that sweeps the problem away in the movement of becoming and redistributes it through the system. As Deleuze and Guattari conclude, "science is haunted not by its own unity but by the plane of reference constituted by all the limits or borders through which it confronts chaos."[91]

In summary, then, along the plane of consistency, there are forms of concepts and conceptual personae. In the plane of composition, there are forces of sensation and aesthetic figures. In the plane of reference, there are the functions of knowledge and partial observers.

TWO BRAINS, THREE PLANES, NO UMBRELLA

Despite the promise of experimentation, the disciplinary boundaries that appear between Zarathustra's qualities and the quantities of Maxwell's demon are not what we have come to expect from a philosophy of mixture. In the relation established between science and philosophy in particular, there is a sharp contrast between the rhizomes that connect royal to nomadic science and the later, more prohibitive encounters with chaos. In the first instance, "what we have ... are two formally different conceptions of science," which are ontologically related to each other in a "single field of interaction."[92] It is here that royal science "continually appropriates the contents of vague or nomad science while nomad science continually cuts the contents of royal science loose."[93] Deleuze and Guattari were certainly, at this stage anyhow, the philosophers of mixture:

> We are always ... brought back to a dissymmetrical necessity to cross from the smooth to the striated, and from the striated to the smooth. If it is true that itinerant geometry and the nomadic number of smooth spaces are a constant inspiration to royal science and striated space, conversely, the metrics of striated spaces [*metrori*] is indispensable for the translation of the strange data of a smooth multiplicity.[94]

In contrast, *What Is Philosophy?* tells us that concepts and functions should only necessarily intersect each other in their full maturity and "not in the process of ... constitution."[95] This is not, as Isabelle Stengers argues, evidence of a nomadic science in the making or a philosophy of mixture but a situation in which each concept and function is created by its own specific means.[96] This separation effectively means that philosophy can only make allusions to science, whereas science can only speak of philosophy as a *cloud* (the Stoic vapor that arises from the actual). Is this really the philosophy of mixture making an almost biblical prohibitive declaration of "thou shall not mix immature creations"?[97] The tripartite structuring of the three planes of science, philosophy, and art, in this way, seems to be a perplexing and shocking defense of what are ostensibly the royal prerogatives of mature science to objective truth. It is as if philosophy and art should hold back from intervening in the creative evolutions of science so as to respect its disciplinary limits. Perhaps *What Is Philosophy?* is emptied of the nomadic voice of Guattari, who, as Stengers goes on to point out, was still trying to escape the limits of scientific functions in *Chaosmosis* published three years after *What Is Philosophy?*[98] To be sure, Guattari's role in the writing of their swansong, while certainly not ghostly, was apparently a tad waspish.[99] But ultimately, *What Is Philosophy?* is not an altogether earnest nod to the autonomy of royal science. It seems more likely that it is a cautious awareness of the need to resist the pressures imposed on each discipline by capitalism. Deleuze and Guattari were indeed writing at a time when there was a lack of resistance to the future—maybe not that dissimilar to the neoliberal attack on the public university we see today. Perhaps they felt that, rather than situate their practice outside of disciplinary borders, philosophy needed to be positioned alongside other threatened practices. The tripartite structure thus becomes an endeavor to view the outside from a point of view founded on each other's internal weaknesses and incapacity to resist the future.[100] These are not so much warring factions as they are disciplines wary of opening up their internal weaknesses to the pressures that can be exerted on them from corporate state power. Given the unremitting neoliberal violence against the public university, as well as the popular industrialization of culture, such distinctions are unlikely to dissipate any time soon. There is in fact a need, I contend, to expand on the planes of *What Is Philosophy?* to include capitalism. Accordingly, Stengers thinks Deleuze and Guattari might have better deployed their own conceptual persona, the idiot, to avoid what is in effect a purification of three distinct lines of creation running from science, philosophy, and art.[101] The idiot need not become embroiled in a contest between rational and irrational public opinion. He would also avoid rushing toward consensual goals. Although he could not mobilize a full force of resistance to business interests, he could at least slow down the pressures. To some extent, then, this book indulges in a certain kind of idiocy in the lecture hall.

My approach in this book closely follows Stengers's cautious reading of *What Is Philosophy?* because it works its way through to a satisfying conclusion in the sense that she supports nomadic interventions into the functions of science without having to adopt the distancing position of social construction or discourse analysis. She concludes by citing the

Deleuze and Guattari of *A Thousand Plateaus,* as I do in what follows, in a slightly extended version:

> Staying stratified—organized, signified, subjected—is not the worst that can happen; the worst that can happen is if you throw the strata into demented or suicidal collapse, which brings them back down on us heavier than ever. This is how it should be done: Lodge yourself on a stratum, experiment with the opportunities it offers, find an advantageous place on it, find potential movements of deterritorialization, possible lines of flight, experience them, produce flow conjunctions here and there, try out continuums of intensities segment by segment, have a small plot of new land at all times. It is through a meticulous relation with the strata that one succeeds in freeing lines of flight, causing conjugated flows to pass and escape.[102]

So here the resistance to the striated spaces of royal science, rooted in hegemonic power, needs to be rearticulated. We are neither respecting the boundaries of scientific functions nor disrespecting the experimentations of others. We instead retain the smooth nomadic spaces in the gaps of royal science. At the same time, though, it is important not to fall into the trap of social construction by disrespecting science, seeing it only as a discursive product of hegemonic power. In the process of becoming part of the strata, science, philosophy, and art all become connected to the apparatus of capture distributed by market power.

CONCLUSION: ENCOUNTERS WITH A THIRD KIND OF INTERFERENCE

> In this submersion it seems that there is extracted from chaos the shadow of the "people to come" in the form that art, but also philosophy and science, summon forth; mass-people, world-people, brain-people, chaos-people.[103]

Despite their discreteness, planes can occupy each other. They can "slip into each other."[104] But as we have discovered, Deleuze and Guattari surprisingly begin by limiting the extent to which things can mix. First, when the concepts of philosophers and sensations of artists become functionalized in neurophilosophy and neuroaesthetics, they become nothing more than functions of knowledge. Likewise, the philosopher only produces a concept of sensation and the artist a sensation of concept. These are the *extrinsic interferences* that occur when, for example, philosophical giants like Deleuze appropriate the musical sensations of Boulez or the mathematical functions of Riemann. Planes meet and resonate, but disciplinary boundaries remain intact so as to negate mixture. Second, though, when Huxley takes the concept of soft control and produces soma and John the Savage, he brings the concept to life. Soma and the Savage become Huxley's aesthetic figures, or rhythmic characters, *intrinsic* to a universe of interference in which concept and sensation slip into each other's plane.

The most promising of interferences, however, appears at the very end of *What Is Philosophy?* Here we meet the indiscernible and *nonlocalized interference* personified by a chaos people to come. "Revolution is absolute deterritorialization," we are told, "even to the point where this calls for a new earth, a new people."[105] For this revolution to occur, though, a complex philosophical maneuver needs to take place. "The philosopher must become nonphilosopher."[106] So what is nonphilosophy? Is it, as Stengers argues, a philosophy that "designates the need for an encounter that does not explain, but produces"?[107] Perhaps it is a philosophy that does not think at all, that only acts. It is only then, when nonphilosophy becomes the earth and people of philosophy, that a new shadow appears—one that seems to overlap all other shadows. At this point, there is indeed a potential shadow of a doubt over the distinctness of the three planes. To be sure, although scientists, philosophers, and artists (standing under their respective torn umbrellas) may well think in distinct ways, they are nevertheless indistinct with regard to the chaos into which they all plunge. Like this, then, the chaotic submersions of the ocean brain become a point of exchange, or a series of traversing lines, through which experimentation can emerge in a nonlocation. This exchange does not necessarily bring about a unifying force between the three planes but rather a junction through which each plane passes on its way, to and fro, chaos. It is at this point of juncture, when the planes are fully submerged, that the concepts, sensations, and functions become

> undecidable, at the same time as philosophy, art, and science become undecidable, as if they shared the same shadow that extends

itself across their different nature and constantly accompanies them.[108]

What is extracted from the chaos is another shadow, that is to say again, the shadow of these people to come of the form that art, but also philosophy and science, summons forth. Is this appeal to nonlocatable science, art, and philosophy a return to the schizoid, the nomad, and the minor in *What Is Philosophy?*[109] Perhaps, just as Stengers lodges herself in the stratum of royal science, in the midst of all those bearded giants of science, the idiotic nonscientific submersions of this book might help to grasp just a little of the potential that is distributed in the actualizations of the neurosciences before it becomes lost to the disciplines of the Schoolman. This is possibly the utility of a nonscientific reading of science. It is also within this differently conceived shadow of neuroculture that we might be able to find an escape route from the apparatus of capture that constitute immaterial neurocapitalism and its alliance with rational science. As Eric Alliez argues, it is within the navel of *What Is Philosophy?* that we find a transformative brain: a *nonobjectifiable* brain that can no longer be treated as a "constituted object of science."[110] It is the nonlocatable assemblages of the ocean brain that defy the model—a third interference or brain, perhaps?

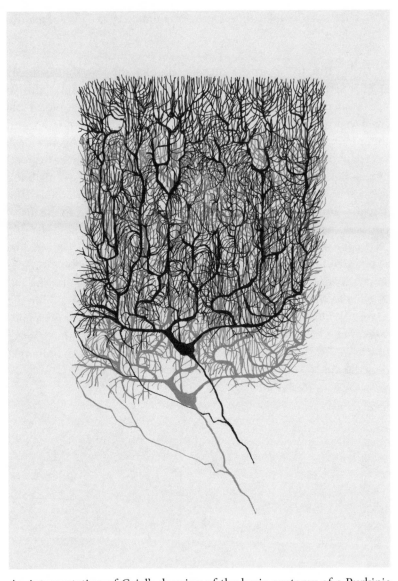

An interpretation of Cajal's drawing of the basic anatomy of a Purkinje neuron. Illustration by Dorota Piekorz.

PART I

What Can Be Done to a Brain?

THE NEURON DOCTRINE

The Spanish pathologist Santiago Ramón y Cajal's (1853–1934) pioneering anatomy of brain cells in the nineteenth century makes him one of the giants of modern neuroscience. It is, nevertheless, Cajal's apparently problematic journey from an inattentive schoolchild obsessed with art to scientific gianthood that produces a number of aesthetic figures that traverse extrinsic and intrinsic interferences between art, science, and philosophy. Not only do Cajal's intricate anatomical drawings lay down the foundations of the neuron doctrine, but his proposition that the brain is made of discontinuous cells would also go on to influence the conceptual basis of the rhizome brain. Moreover, his early childhood struggles with paying attention to authority at home and in the schoolroom, and later efforts at writing dystopian science fiction, mark him (and his aesthetic figures) out as an intrinsic interference that brings to life a society of neuroscientific control. He is, like this, an interesting catalyst for both tracing the early trajectories of neuroculture and framing the question concerning *what can be done to a brain.*

ANATOMY, FUNCTION, AND INTERVENTION

Cajal's influential anatomical work in the late 1800s lays the foundation of modern neuroscience and opens up two significant research trajectories. On one hand, this giant observed, and exquisitely sketched, the composition of nerve cells. On the other, he pondered the processes underlying the mediation of brain functions through these cell populations.

Cajal's work was complemented by contemporary improvements in microscopy and staining techniques, particularly a method developed by the Italian physician Camillo Golgi in the late 1890s. This involved a painstaking, lengthy process of exposing brain tissue to dichromate salts before impregnating it with ions made from silver or mercury, making it possible to see the various components that compose the nervous system in a full panoramic view, including the cell body, dendrites, dendritic substructures (spines), and at least part of the axon. Cajal would apparently observe the stained tissue under his microscope for long stretches of time before drawing the cells from memory with a glass of absinthe.

Gradual improvements to the Golgi method led Cajal to conclude that the brain was not, as the dominant reticularist theory had claimed, a continuous meshwork of cytoplasm (a reticulum). Following Cajal's discovery of dendrite spines in the cerebral cortex in 1891, the neuronists argued that the brain is instead made up of discontinuous cells confined to contiguous, autonomous individual neurons: the "butterflies of the soul," as Cajal referred to them.[1] At first, the reticularists rejected the authenticity of Cajal's dendritic spines, claiming that they were artifacts (or noise) produced by the Golgi method rather than genuine structures. They also tried to explain them away as sites of cytoplasmic continuity between dendrites (a form of dendro–dendrite connectivity). However, Cajal set about supporting the theory of discontinuity and rounding on his reticularist opponents, describing their notion of continuity as a "chimera"—a "contrived house of cards," this "panreticularism [is] absolutely unobservable."[2]

Along with Golgi, Kölliker, His, and other giants, Cajal was eventually reviewed in a series of papers published by Wilhelm von Waldeyer-Hartz, also in 1891, which summarized these new findings in a coherent theory. Waldeyer-Hartz concluded that "specific types of nervous system cells, called neurons or nerve cells, constitute the anatomical, physiological, metabolic and genetic unit of the nervous system."[3] So it was Waldeyer-Hartz, not Cajal, who first coined the term *neuron,* and he went on to be credited with establishing the neuron doctrine, that is to say, the proposition that individual nerve cells communicate at regions of cell-to-cell *contact* where there is no continuity between cells.

Following the neuron doctrine, the research focus shifted to the underlying processes of communication between nerve cells. By 1897, the British neurophysiologist Charles Sherrington had pointed to a possible anatomical and functional solution to Cajal's as yet unproven theory of dynamic polarization. Despite the morphological data Cajal presented, which pointed toward the "existence of synapses and the one-way property of neural data transmission[,] he was not able to realize the intraneuronal reversibility that Sherrington [using myographic recordings of the spinal reflex system of jellyfish, cats, and dogs in his lab in Oxford] proved experimentally."[4] In 1906, Sherrington published *The Integrative Action of the Nervous System,* in which he first theorized about the gap between neurons, calling it the synapse (taken from the Latin for "to clasp"), and pointed to the crucial interplay of excitation and inhibition in brain functioning.[5]

Working out how the synapse functioned became the overriding challenge for early neurophysiology. Indeed, before the 1930s, many neuroscientists, including Sherrington, regarded synaptic events to be electrical. Nonetheless, in 1921, Otto Loewi, a professor of physiology from Vienna, established an experiment demonstrating what he termed neurohumoral transmission. That is to say, Loewi produced inhibition and acceleration in hearts removed from frogs, revealing underlying chemical processes.

At this point, the trajectory of the neurosciences seemed to bifurcate in two directions, both of which would have a wider influence on life outside of the lab. First, efforts made to map neurophysiological functions to psychological states have drawn attention to the potential control of subject behavior by way of relating physiological brain matter to psychological and behavioral states. This line of flight initially became apparent in Ivan Pavlov's objective psychology and the influence it exerted on nascent Spanish brain science.[6] Even before Pavlov arrived in Madrid in 1903 to talk to the XIV International Congress of Medicine about "psychological reflexes," Spanish scientists had sought to remove the barriers separating the two. As Martín Salazar noted in 1880, although the deep-seated differences between organic and inorganic matter had been "torn into a thousand pieces by modern biological studies,"[7] there was a continuance of another barrier that separated psychology and physiology. This was a barrier that isolated the moral world from the material world. Yet, as Salazar notes, "ever since the cerebral reflex action was demonstrated . . . from this very moment, the secular wall has fallen down."[8]

Although a tardy Pavlov apparently failed to make the submission deadline for the congress proceedings, it was his theorization of the reflex system, and its apparent capacity to learn by conditioning, that eventually provided early brain science with a behavioral hypothesis to accompany Cajal's anatomical findings. In fact, Cajal was among a number of neuroscientists who attended Pavlov's talk who became interested in objective psychology and its account of conscious phenomena in terms of cellular occurrences governed primarily by mental association mechanisms.[9]

The second strand of neuroscientific research has focused on the functioning of chemical processes that occur between nerve cells, or neurotransmission, as it is now called. The possibility of direct chemical interventions into conscious and unconscious brain processes, assumed to affect mental and emotional states, looms large. Certainly this is the point at which the question concerning *what can be done to a brain* is approached through invasive experimentation at the level of neuronal interactions. Moreover, potential chemical treatments for mental disorders like depression are counterpoised with what the neuroscientist Steven Rose calls the Huxleyesque implications of chemical interventions into the brain.[10] As soon as Loewi identified the role of chemicals in neurotransmission, the potential for neuropharmacological manipulation of brain states became evident. By the 1950s, synthetic chemical interventions had been deployed to increase or decrease the strength of communication across a synapse. Indeed, much of this neurochemical research has been used to inform the business of the pharmaceutical industry, where the functions of the synaptic event have been exposed to increasingly sophisticated neurochemical products. Today, drugs like Prozac, Provigil (modafinil), and Ritalin, all of which target neurotransmitters, are designed to affect mental states associated with happiness, wakefulness, and attention. To be sure, as Rose contends, these cognitive enhancers have the potential to become the soma of the twenty-first century.[11]

CAJAL'S ATTENTION DEFICIT

Cajal's journey from artistic child to giant of neuroscience is a difficult one, complicated by a complete lack of interest in school and a seemingly compulsive obsession with drawing that infuriated his father. By his own account, as a young boy, he aspired to be an artist, but his father (a medical doctor) was having none of it and fiercely opposed the young Cajal's ambition. How could he possibly make a living from drawing? Despite this, his studies were of little interest to the young Cajal. "His attention always wandered and his hand had to doodle."[12] Cajal was eventually packed off to a school with a brutal educational philosophy of *la letra con sangre entra* (knowledge enters with pain). But he had no aptitude for strict rote learning and became a disruptive force in the classroom. He was apparently "untamable"—a lost cause—and would often escape into the countryside to draw. As a punishment for his misbehavior and inattentiveness,[13] the disobedient Cajal was treated to a reign of terror. He was whipped, starved, and locked in a dark room by the friars who ran the school.

Today, in times of neuroculture, Cajal would of course be treated very differently. He would still be, initially, disciplined for this behavior (perhaps not so violently, one would hope), but his short attention span, sustained bouts of distractibility, hyperactivity, and impulsiveness would most probably lead his teacher and parents to refer him to the family doctor, who might, in turn, refer him to a psychologist, a psychiatrist, and, eventually, a neurologist for electroencephalography (EEG) testing. Indeed, if Cajal's behavior corresponds with a certain higher ratio between two brain waves than is the norm, his condition might eventually be attributed to a problem with his brain. In short, Cajal would be diagnosed, like an increasing number of children in the United States and the United Kingdom, including those at primary school, with ADHD.[14]

Cajal eventually dropped out of school altogether and managed briefly to attend an art school in Aragon, but his father never recoiled, and the would-be artist was ultimately forced to study medicine. In his memoirs, he remembers the bitterness of his defeat: "I must exchange the magic palette of the painter for the nasty and prosaic bag of surgical instruments!"[15] Despite his earlier bouts of inattention and hyperactivity in the schoolroom, he eventually reoriented his obsession for art and became captivated by anatomy, microscopes, and the "life of the infinitely small."[16]

FOR THE ATTENTION OF DR. BACTERIA

Although his work focused primarily on anatomy and function, Cajal was, it appears, fascinated by the question of what can be done to a brain. It is here that his interest in hypnosis, Pavlov, and dystopian science fiction unleashes intrinsic interferences. It is indeed Cajal, the artist, not the scientist, who emerges, later in life, to conceivably bring together the neurosciences and a proto-Huxleyesque concern with the powerful hypnotic suggestibility apparent in Pavlovian conditioning. In his 1905 collection of short science fiction tales, Cajal (writing under the pseudonym Dr. Bacteria) introduces an outlandish aesthetic figure of his own: the bourgeois Dr. Alejandro Mirahonda, who lives in times of early industrialization, beset with aristocratic fears of violent class struggle. Mirahonda is a great hypnotizer, clever enough to realize the boundaries of his own magnetizing powers over the thieving, brawling, and drunken behavior of the newly industrialized proletarian. How then to overcome such antisocial passions as the idleness, rebelliousness, lasciviousness, and criminal instincts of the poor? Mirahonda's answer is, in part, to move away from the blood and pain of class warfare carried out on the streets and in the fields toward a mode of soft power founded in the factory, the scientist's lab, and the sociologist's study.[17] But before such remote ideals can be fully realized, suggestibility needs to be readily deployed to pacify those who "lack good brains or good fortunes."[18] To begin with, Mirahonda argues that the proletariat need to be *got at* from a very early age, their plastic cerebral centers placed under unremitting bouts of social hypnosis. But for reasons both Dr. Bacteria and Huxley similarly grasped, these hypnopedagogic programs have their limits. Mirahonda explains:

> In a group of a hundred people, selected at random, only fourteen or sixteen will be capable of being hypnotized and suffering, by way of suggestion, amnesias, paralyses, contractions, emotional transformations, hallucinations, etc. A very prestigious hypnotist, who knows how to strike deeply at the public's imagination, can increase this figure to twenty-four, maybe thirty. But despite all his efforts, he'll still be left with the remaining seventy percent:

inattentive, open-minded, refractory to any belief in miracles, and therefore immune to suggestion.[19]

In the short term, the powers of suggestion require a conditioned stimulus, but this must go beyond the old behaviorist game of repeating the stimulus until it becomes a habit. The stimulus needs to be concealed, hidden under the white coat of a scientist or the cape of a saint. In short, Mirahonda's hypnotic suggestion is accompanied by a placebo in the shape of a sugarcoated suggestive pill, which simultaneously taps into the anticipations of the populace, drawing attention away from the trickery of the hypnosis itself, and fulfills the objective of absolute social docility. In the long term, Mirahonda's seemingly benevolent concern for the poor offers nothing short of a scientific purification and sublimation of the proletariat.

2

Neurolabor: Digital Work and Consumption

In this chapter, I pursue the question of what can be done to a brain by mapping the brain's relation to recent trends in the management of efficiency in the workplace and sites of consumption. This tracing exercise will be set against the backdrop of an ostensibly familiar discourse concerning a shift from Taylorist to post-Taylorist factory models, that is to say, ongoing efficiency analysis caught in a transition from material to immaterial labor. The subsequent discussion is organized around three paradigms of computer work that are reappropriated from the study of HCI. The aim is to draw attention to the ways in which the everyday lives of workers and consumers have converged into a complex, circuitous, and exploitative mode of capitalism that increasingly makes use of the brain sciences to root out inefficiencies. It must be noted that this is a distinctly political rendition of HCI, a discipline not usually renowned for its critical interventions into unscrupulous modes of capitalist efficiency, analysis, and management.[1] In fact, my point is that HCI is more often than not complicit in initiatives directly linked to Taylorism. This chapter observes, as such, the many continuities and discontinuities associated with a shift from the muscular, rhythmic entrainments of industrialized labor (analyzed according to ergonomic, social, and psychological factors) to the introduction of cognitive and, more recently, neurological models, which have coincided with the digitalization of work and consumption and, simultaneously, drawn both into a circuitry of control.

In short, I argue that although there is a considerable shifting of ground, mainly brought about by changes in technology and scientific approaches to brain–body coupling, the goals of the efficiency management of work and consumption (to combat the evils of inefficiency and conform bodies, minds, and brains to the quickening rhythm of capitalism) remain consistent. Indeed, as ergonomics and cognitive science give ground to the neurosciences, and digital technology becomes increasingly ubiquitous, the efforts made to exorcise inefficiency and nonconformity from the workplace and sites of consumption also become more intensified. This is indeed a manifestation of neurocapitalism in which neuronal interactions, assumed to relate to emotions, affect, feelings, and decision-making processes, are put to work in the fight against inefficiency.

The chapter concludes by grasping these indices of neurocapitalism that transpire in HCI through the concept of *ersatz experiences*, that is, part of a regime of control that neurologically models and imitates the felt experiences of everyday life and reworks and recycles them to condition such things as worker motivation and consumer engagement. What this amounts to is a deepening of a technological unconscious and increasing control of the sensory environments in which people work and consume. It is indeed this reworking of experience that currently melds together the worker and the consumer in the same circuit of control.

FROM GRAMSCI'S BRAIN TO THE CYBERNETIC FACTORY: TWO CAMPS

Although cerebral labor is certainly not a newcomer on the factory floor, the digital reorganization of work in the late twentieth century has significantly intensified the open circuitry that connects the brain to postindustrial working life. The brain has, as such, been more finely tuned in to the rhythmic frequencies of what has been called cognitive capitalism. Aspects of this intensification have already been well documented in diverse literature covering the shift from manual to so-called cognitive labor. To begin with, these various approaches can be crudely located in two camps. On one hand, there are numerous popular notions of the *smart* advantages achieved through sharing knowledge on a network. These serve as indicators of the emergence of an assumed neoliberal economic model that taps in to the self-organizing hive mind. On the other, the detriments of immaterial labor have been grasped as part of a cybernetic control system that transforms activities not normally associated with work (play, chat, fashion, tastes, opinion) into products of often low-paid or even free labor.[2] Indeed, the notion of emergent collective intelligence becomes something that is exploited by the market rather than celebrated as an empowering force that might lead to mass nonconformity and potentially become an enemy of capitalism.

There are nevertheless complex subtleties apparent in this second camp that need to be cursorily acknowledged before progressing. As Terranova argues, although often portrayed in much of the Anglophone world as a crude shift from factory to information labor, the exploitation of factory labor has never gone away. The debate among Italian post-Workerists and French post-Marxist philosophers about what has changed is indeed far more complex than is often depicted.[3] For example, commenting on the work of Maurizio Lazzarato, Terranova points out that the shift he describes

> under the heading of immaterial labour [cannot] be dismissed as a simple quantitative transfer of surplus value from the factory to the "upper" floors of capitalist production. What it indicates is that the core of production now directly concerns the production of subjectivity: affects, desires, beliefs, aspirations, knowledges, ways of living.[4]

The focus in this chapter therefore aims to provide an extension of this latter approach, which specifically takes into account a series of shifts in the neurological models that inform ideas about brain–body relations in the digital workplace. Certainly the catalytic circuitry of neurocapitalism presented here must not be mistaken for a benevolent collective brain trust or an awakening of a collective cognitive consciousness. It is, rather, a mode of subjectivity in the making that not only overloads and speeds up the time of brain–body relations in the workplace but also exploits the desires, beliefs, emotions, feelings, and affective states of the worker to a point where the capacity to think independently is diminished and, worse, people drift away from nonconformist states of mind necessary to resist progressively more intensified frequency-following, routinized, and debasing working conditions.

Importantly, in times of neurocapitalism, brain labor is not simply distinguished from manual labor in the sense of the former being all about cognitive, conceptual, and nonempirical work, while the latter is all about the expenditure of physical energy. We need to further consider noncognitive components of labor. To be sure, the circuits that connect the brain–body relation to the workplace are also conditioned by managerial efficiency drives, aimed directly, and indirectly, at brain-body functions; associated with sensations, feelings, emotions, and affect; and tending to circumvent cognitive processes altogether. Indeed, today there seems to be more and more effort made to put neuronal interactions to work below the threshold of conscious cognizance. Nonetheless, this is not to say that work (or consumption) becomes wholly unconscious or, indeed, unthinking but, more exactly, affective states develop a mind of their own that more readily conforms to the punishing work rhythms of capitalism.

GRAMSCI'S BRAIN TAKES A WANDER ON THE FACTORY FLOOR

Albeit following a familiar trajectory of efficiency-driven initiatives already applied to the entrainment of bodies and brains in the industrial workplace, there is at the same time something equally novel about these recent attempts to couple neuronal interactions to labor processes. To fully explore these complex continuities and discontinuities, the discussion begins by tracing back the ongoing transformations of the brain–work relation to the early days of the twentieth-century mechanization of trades, where, surprisingly perhaps, the brain was considered to be liberated from muscular labor. As Antonio Gramsci argues—in his analysis of Americanism and Fordism—the tradesperson is modified by the scientifically managed assembly line just as a child is adapted when he learns to walk. Like the child, the worker's physical motion becomes automated and her memory is reduced to "simple gestures repeated at an intense rhythm, 'nestle[d]' in the muscular and nervous centers."[5] Significantly though, the memory of the trade is grasped by Gramsci not in terms of cognitive memory but rather as a kind of muscular *habituation*. Once a child has learned to walk, he can do so without thinking. Because thinking becomes surplus to use in the act of walking, there is a kind of emancipation of the brain. "One walks automatically, and at the same time thinks about whatever one chooses."[6] So, when the industrial adaptation of the tradesperson is complete, we do not find a worker with a *mummified* brain. Far from it: the brain "reaches a state of complete freedom," unfettered for other preoccupations.[7] This is, of course, a limited kind of freedom, but it could lead to the realization that in effect the Fordist factory model reduces the worker to nothing more than what Gramsci called a *trained gorilla*.[8] Although clearly such freedom to think does not (1) constitute absolute freedom from factory discipline or (2) help us grasp the role of habitual and cognitive memory in the workplace, it does nonetheless have the potential to lead, as Gramsci notes, to what industrialists would consider dangerous thoughts of nonconformity.

The onset of cognitive capitalism is perhaps anticipated at this point by what Gramsci identifies as the industrialists' concern with the threat posed by nonconformity resulting in a series of cautionary measures and *educative* initiatives well evidenced in Henry Ford's early experimentations with trade schools that brought together prescriptive modes of academic study and industrial instruction into one syllabus. Although, during this period of time, the Fordist factory continues to be a site of material production, these educational initiatives usher in a cognitive subjectification of the worker in terms of conditioning individual knowledge while, at the same time, ensuring proprietary skill sets and priming competitiveness in the workplace. As Ford puts it, "the man who has the largest capacity for work and thought is the man who is bound to succeed."[9] Indeed, Ford recognizes how modern systems of work required "more brains for [their] operation than did the old [systems]"; there was a need, he contended, for "better brains" to run the "mental power-plant."[10] This early example of industrialized education, focused on cognitive and manual labor, is perhaps an early indication of what would eventually become known as immaterial labor.

IMMATERIAL LABOR?

In the post-Fordist era, the notion of immaterial labor needs to be approached carefully. It has seemingly developed around the accumulation of mostly intangible goods, namely, information- or knowledge-based products, such as software, but also pharmaceuticals and genetically modified agriculture. Likewise, cognitive labor is (1) supposed to be increasingly organized around immaterial skills and activities associated with brainpower, including attention, perception, and memory, and (2) distributed through nonlinear information networks, which bring cognitive subjects into productive and competitive relation with each other. However, significantly, what has also changed in the wake of the continuous computerization and cognitivization of the workplace must not be mistaken for a generalized trend toward creative, nonroutinized knowledge work. The so-called rise of the *knowledge worker,* free from the routinized drudgery of the assembly line, needs to be seen as part and parcel of a myth created by the advocates of a new economy model intended to intensify efficiency management. Creative brain time required for new ideas, experiences, interpretations, judgments, and inventions may well be in demand in privileged regions of the world, such as Silicon Valley, but job descriptions like those associated with software engineering are of course vastly outnumbered by highly routinized, low-skill work in call centers and online retail warehouses. The need for highly skilled cognitive labor is evidently inconsistent with a global digital economy overwhelmingly dominated by a manual workforce engaged in low-skill interactions with computerized factories.

The cognitive subjectification of the post-Fordist worker was clearly never intended to encourage the freedom to think outside of the cultural circuits of capitalism. To be sure, even when creative brain time is in demand, it has become increasingly routinized by computer systems that churn out creativity and innovation as a kind of habitualization of the labor process, not unlike Gramsci's conditioning of muscular memory. In many ways, then, cognitive capitalism conforms both low-skilled, practical and creative thinking to a flexible production process. This is a network of labor that connects increasingly mobile and adaptable digital factory spaces to a globally deregulated and competitive (cheap) workforce, while also fervently using automated online software to disintermediate the middle space that once stood between the high street retail outlet and the consumer purchase.

NEW WEAPONS

Cognitive capitalism can also be conceptually grasped according to a rigid cybernetic model of the brain, transforming Gramsci's muscular rhythms of industrial work into a mode of immaterial labor consistent with the computer–mind metaphor of cognitive science. Cognitive routinization is therefore perfected by way of conforming mind processes akin to software, including perceptive and attentive functions, to inputs consigned to a memory storage or hardware system primed for action (work). However, just as there are practical limitations to the implementation of a purely cognitive workforce, there are conceptual limits to the robustness of this cybernetic model of the mind. In short, although the computer–mind metaphor becomes central to the study of cognitive digital labor, it marginalizes bodily interactions (muscular and affective) with computing that are directly related to cognition. In fact, as a model capitalist subjectification of the worker, the cognitive subject is flawed in many ways. My point is that, similar to Gramsci's brain in the Fordist factory, cognitive circuits of control have not been able to completely mummify the capacity to think nonconformist thoughts that might once have led to strikes and industrial sabotage. Indeed, within the coupling of digital networks and cognitive computer-minds, dangerous thoughts of nonconformity have managed, albeit in a narrow sense, to persist. As Deleuze profoundly argues in his influential and significant control society thesis, the digital circuits of cognitive capitalism would indeed prove to be prone to thwarting new weapons that introduce new modes of instability.[11] The potential of hacking, virus writing, denial of service attacks, digital piracy, WikiLeaks, and virtual petitions and occupations bolster, to some extent, nonconformist social movements online, which, although limited in their resistance to vigorous efficiency management, provide at least some evidence of latent disorder in digital circuitry. There is always a desire for new weapons.

NEUROLABOR AS EXPERIENCE

Although the full charge of cognitive capitalism is far from exhausted, it is important to note how developments in the neurosciences, particularly those involving an apparent deeper understanding of emotional and affective brain–somatic relations, are supplementing efficiency analysis and management. Managerial control has, it seems, switched its attention away from the software–hardware dichotomy of the immaterial mind in favor of indirect access to hardwired material brain functions. To be sure, the business enterprise has been quick to realize the potential utility of neuroscientifically inspired ideas, for example, in effecting "change management," encouraging "compassionate" corporate communication, and working with big neuroscientific ideas, such as neuroplasticity, to create a "brain-friendly workplace" in challenging economic times.[12] The business enterprise blogosphere is currently awash with neuroscience-enthused managerial techniques concerning motivation, emotional intelligence, creativity, and workplace mindfulness.

This steady production of efficient brain labor has coincided, to some extent, with what Bernard Stiegler has identified elsewhere as *neuropower*: a shift away from the biopower of the factory floor (and schoolroom), and the psychopower of marketing control, toward the creation of new markets for consumption.[13] There certainly seems to be a marked effort by the capitalist corporation to tap into the synapogentic processes of digital workers and consumers. However, it would also appear that the efficiency drives that underlie biopower, psychopower, and neuropower (the forces that control bodies, minds, and brains) have not simply been usurping each other. They have instead become even more interwoven in the cultural circuits of capitalism and its ongoing efficiency analysis. Indeed, the eventual capture of Gramsci's brain in these circuits, can, as such, be grasped here as part of an unremitting circuitry of control in which any inclination toward a nonconformist brain is increasingly confronted, exorcized, and substituted by the production of brain-friendly environments, affective atmospheres, and ersatz experiences, that is to say, a postindustrial layering of mostly artificial experiences related increasingly to brain functions associated with emotions, feelings, and affective states, in addition to muscular movements and cognitive inputs and outputs of previous industrial factory models.

As follows, neurolabor might be further conceptualized as an expansion of cognitive capitalism, typified by the attention economy model, into a noncognitive mode of capitalism, outlined, to some extent, in new economic models focused on experience.[14] We might also link this expansion to a mode of neurocapitalism, because it is the control of the sensory environment that sets the rhythm of working life and consumption that really matters. To be sure, the experience economy reconfigures the relation the computer worker and consumer have with the tangible and intangible production of commodities by way of adding to and reworking previously felt experiences. Considered as a type of post-Fordist factory, the production of these reworked, or ersatz, experiences more readily compares to a refinery or distiller model than it does to assembly-line production. This is to say, production takes the raw ingredients of previously felt experiences, like those relating to compassion, fun, or personal value, for instance, and modifies them into novel inventions that encourage more efficient production and consumption by ramping up emotional engagement. The ersatz experience is in fact more Walt Disney than it is Henry Ford,[15] or indeed, following Huxley's dystopia, we might substitute the term *post-Fordism* for *transcendental Fordism*: God of the assembly line, immaterial and experience labor.

The production of ersatz experiences becomes apparent in recent developments in HCI research and in subsequent trends in interaction design, including UX design. Here neuroscientific ideas concerning the emotional brain's role in decision-making processes relating to navigational choices and mouse clicks, for example, become indispensable to the efficiency management of computer work and consumption expressed through design, branding, and marketing.[16]

THE ROLLING PARADIGMS OF HCI

It is possible to trace the origins of neurolabor back through historical shifts in the management of computer work and consumption. Indeed, these shifts can be located in a trajectory of research identified elsewhere as the so-called *three paradigms of HCI,* which, before going on to "document underlying forces that constitute a third wave in HCI," set out two "intellectual waves that have formed the field."[17] In this section, I will call these first two waves *ergonomic* and *cognitive,* which, respectively, stem from "engineering/human factors with its focus on optimizing man-machine fit," on one hand, and an "increased emphasis on theory and on what is happening not only in the computer but, simultaneously, in the human mind," on the other.[18] The point is that these initial shifts in HCI research not only follow the transition from the Fordist to the post-Fordist factory model but also point toward a growing focus on *user experience HCI,* inspired, to some extent, by a combination of marketing and neuroscientific ideas concerning the emotionality of the brain.

Before I approach these paradigm shifts, I think it necessary to set out a series of theoretical preconditions, because mixed in with the constant paradigmatic change are a number of invariant political components that need to be accounted for:

1. The striving for Fordist and Tayloristic managerial efficiency drives remains invariable in each paradigm. What has changed is that techniques of efficiency management have been rolled out variously throughout each paradigm, affecting ergonomic bodies, cognitive minds, and emotional and feely experiences, as a kind of apparatus of capture of labor. The production of efficient bodies and minds can be accounted for by way of existing labor theory, but the latter focus on emotional experience requires, I contend, a new theory of neurolabor. To be sure, such a theory needs to include neuroscientific components, which can help explain why, for example, the neuroscientist Antonio Damasio's work has had such a profound influence on experience-processing models deployed in the UX industry.[19]

2. Even if individual components move around in the circuitry of each paradigm, the factory model, of the kind Gramsci's brain first encountered, has gained some level of fixity. Although such things as homework, mobile work, and, significantly, the work of the consumer as coproducer provide evidence of free-moving components in the capitalist circuitry, the rigid exploitations experienced in the Mexican *maquilas* or the Amazon factory model, for example, remain invariant. Certainly, as HCI practices shift further away from a focus on purely cognitive user interaction in the workplace toward affective computing and consumption, we are likely to see an intensification of worker conformity as well as subsequent novel struggles for nonconformity.

3. We need to consider carefully what it is that constitutes paradigmatic change. To this extent, my approach differs from that of *The Three Paradigms of HCI*, which, although acknowledging a sometimes noncontradictory shifting of theories and practices, concentrates more on a clash of metaphors occurring at the center

of each distinct paradigm. In this reading of transformations in HCI research, I am as interested in looking for invariant singularities as I am in the variant features of each paradigm. I will therefore approach each paradigm as an open system rather than as a closed or discrete entity. Moreover, in times of unpredictable technological change, the term *paradigm shift* has been frivolously applied to seemingly distinct and autonomous emergences (old and new). It is therefore essential to be reminded of a more complex set of diachronic emergences like those Thomas Kuhn first attributed to scientific paradigms.[20] There is, for instance, no higher deterministic authority able to capture and guide the trajectory of science. A paradigm shift is, it would appear, emergence with no downward causation. Each shift is defined by what escapes it, namely, the scientists who jump ship from normal to new science. What is more, in this analysis of the shifts assumed to be occurring in HCI, it is necessary to tread carefully by noting that a paradigm is a combination of mostly unpredictable variation alongside a gentle rolling out of undulating extensive invariance. The first ergonomic paradigm (from the Latin *ergo,* for "work," and *nomos,* for "natural laws") is, for example, a combination of physiology, psychology, and social factors that do not disappear in the ensuing paradigms but persist or reemerge in slightly different contexts. The third paradigm is similarly endemic to an enduring control circuitry that brings the territorializing forces of ubiquitous computing, emotion research, the utterances of design gurus, UX consultants, and the strategic business school inventions of the experience economy into the otherwise shifting assemblages of capitalism.[21] Each paradigm of HCI is, as such, an emergence of metric extensive properties underscored by immeasurable intensive differences—a kind of moving juxtaposition.

4. It is also important to observe a persisting trend in the natural laws of work, particularly in their manifestations in HCI, toward an accumulation of disciplinary approaches, which seems to pick up speed and mass through each paradigm shift. Initially marked out as a coming together of physiology, psychology, and, later, social factors, HCI has acquisitively expanded its reach to

encompass sociology, cognitive science, computer science, anthropology, and industrial design, and now, in the so-called third paradigm, it is the neurosciences that are being drawn into its maelstrom to further proliferate the laws of work. On one hand, then, we have a qualitative grasping of indivisible cognition, emotion, and feeling and, on the other, a quantifiable rolling out of efficient and divisible bodies, brains, and machines.

THE ERGONOMIC PARADIGM

Efficient Bodies

The first paradigm of HCI can be defined as an "amalgam of engineering and human factors" in which interaction is grasped as a physical coupling of human and machine, the goal being to optimize the best fit between the two. Here we enter into the pragmatic world of industrial engineering inspired by Taylor's scientific management techniques already established in the early part of the 1900s but gradually expanding into more complex machine systems and early forms of computing. The objectives of Taylorism remain constant throughout this period, that is to say, to counteract the evil of inefficiency and increase worker effectiveness by way of the introduction of two major innovations: first, the division of labor according to managers and workforce, and second, the breaking down of the physical movements of the worker according to time and motion.

Social Factors and Beyond

Unlike Taylorism, though, the first paradigm is not defined by the labor of the body in isolation. It has its origins in war, specifically in the military use of advanced human–machine systems in World War II, which ushered in a new wave of technological invention requiring brains as well as brawn. After the war, the physical and mental demands of these human-operated machines found their way into the state and commercial industrial sector, requiring a new kind of managerial approach supported by military and academic research into physiological and psychological labor. Indeed, by 1949, the British Admiralty had proposed the name "ergonomics" (the natural laws of work) to describe a new kind of discipline concerned with the physical factors of work, but some three years later, the Ergonomic Society was formed, employing people from psychology, biology, physiology, and design to assist in developing this new approach to work. In the United States, too, psychological factors became of increasing importance in the study of post–World War II workplace efficiency. The Human Factors Society, formed in 1957, focused on the social role individuals played within complex industrial systems.[22] These early endeavors to bring together ergonomics and social factors permeated the working milieux of the latter part of the twentieth century in a number of ways.

First, efficiency analysis focuses on worker movement through the factory space. There are evaluations and measurements of a series of demands on the worker, including physical demands required to lift and move material objects through a space and skills necessary to operate machines to schedule. This focus on making bodies move more efficiently through the workplace comprises a consideration of bodily dimensions, competences, and physiological processes. More significantly, it concentrates on making working processes error or accident free. Indeed, error analysis leads to more reliable systems, making them, on one hand, easier to use, more comfortable, less fatiguing, and less stressful and, on the other, more profitable.

Second, each paradigm seems to have its own unique pathologies. Like this, in the ergonomic paradigm, a worker's material and immaterial interactions with tools, furniture, heat, noise, vibration, and pollutants in the industrial factory are taken into account. For instance, vibrations can be transmitted to a human body through contact with external vibrating surfaces, such as a handle or seat of a machine, making the body oscillate to the rhythm of the workplace environment, which is of course endemic to most industrial labor environments. However, certain extreme interactions with vibrating machinery are classified according to damaging whole-body or hand–arm vibrations, which can lead to conditions like white finger, resulting in long periods of inefficiency due to time off work.

Third, and together with the advent of computing technology, workers are measured according to how they move through the virtual spaces of the digital factory. Tool design processes are coordinated according to anthropometric and biomechanical data captured directly from HCI, including task-based analysis involving hardware controls and displays and software graphical user interfaces. This is the origin of HCI-based user testing research into the efficient interactions of computer work, including the measurement of tasks consistent with the number of clicks made while navigating through a software system, for example.

Finally, the ergonomic paradigm begins to focus on the social, cultural, and psychological conditioning of the worker. It initially draws on primarily behaviorist methods, such as the pace of work and training, but progressively relates analysis to the impact of collaborative work, mental workload, and information processing, in addition to a consideration of how worker motivations can be factored in to the study of labor efficiency.

Gramsci's Brain Becoming Digital

Returning to Gramsci's unencumbered brain—free to think nonconformist thoughts—we begin to see how such freedoms are still obtainable but gradually eroded by the intensification of efficiency analysis ushered in by the onset of digitalized labor. There is initially enough residual brain power to perhaps grasp the exploitative conditions of repetitive mechanical work and imagine a different future, but the routinization, fragmentation, and cumulative workload of digital labor put pressure on the brain time necessary to resist. Certainly, in the era of large-scale factory assembly lines, an organized workforce could slow down or remove physical labor from the production process. Physical force could also be used to sabotage machinery. However, coinciding with the large-scale industrial unrest of the 1970s, resulting in new curbs on union power, new digital technologies have provided employers with the wherewithal to reorganize the workforce along more distributed lines of production, while, at the same time, ergonomics has branched out from physical and social factors to encompass an increasing focus on the cognitive processes of the worker. The time and space necessary for Gramsci's brain even to contemplate, let alone plan and execute, modes of resistance in the workplace are gradually compressed.

THE COGNITIVE PARADIGM

Efficient Minds

The second paradigm needs to be seen against two shifting backdrops. On one hand, well-documented and vivid technological modifications to the assembly-line model introduce a progressively more flexible computerized production process. On the other, there is an increasing focus on psychosocial and cognitive aspects of labor realized in the advent of HCI. Although quite often embellished rhetorically as focusing on aspects of *user need,* HCI at this initial point is, arguably, a technosocial expansion of Taylorism still focused on combating the evils of inefficiency but moving away from the worker grasped as a cog in a machine to a worker coupled to the machine as an information processor. Indeed, the inception of the digital factory and its networked proliferation of routinized cognitive labor into all areas of life introduces new opportunities to analyze and manage physical and psychological labor inefficiency on-site and remotely. In the first instance, worker movement becomes even more unitized, observable, flexible, and distributed throughout the digital factory space (and beyond), making the individual ever more remote from Gramsci's desire for collective nonconformity. Various graduations of outsourcing labor processes across national regions and international borders produce fewer opportunities to collectively congregate, protest, or strike effectively in single locations. The network surveillance of worker movement is also extended beyond the factory space to locations previously outside of managerial control, including the journey to and from work, the home, and even the vacation. To be sure, computer technologies, which were once only found in the workspace, begin to spill over into domestic and leisure spaces, blurring the distinction between worker, user, and consumer.

From the 1970s onward, the influence of both the ergonomic concentration on psychosocial factors and new insights from cognitive psychology converge in the discipline of cognitive HCI. Harrison et al. grasp this second paradigm as a different kind of human–machine coupling to that expressed in early ergonomics. Computer work, they contend, is now increasingly defined and organized around "a central metaphor of mind and computer as symmetric, coupled information processors."[23] Managerial strategies become concentrated on the design of the interactions that facilitate the rapid information flow between processors (users and computers). Information flows in and out of this coupling of processors, undergoing transformations as it passes through, which can be manipulated so as to ensure more efficiently communicated flows.

The Computer Mind Goes to Work

The second paradigm marks the emergence of the cognitive subject as defined by a new instrument of labor: the computer mind. The initial aim, it would seem, was to encapsulate digital labor in an information model founded on mostly crude analogies between the seemingly unidirectional and sequential processors of the black-box mind and a computer. A worker would, for instance, encode information received from the environment as an internal representation stored somewhere in the mind and selected for comparison with other stored representations. This flow of information would act as an encoded stimulus that would trigger the organization and execution of a response as an action. These early cognitive models would later be expanded to include perception, attention, and memory, opening up a series of further questions relating to how information is perceived by perceptual processors, attended to, and stored in sensory short- and long-term memory. Indeed, the worker's *mental model* is divided up like a computer memory in the sense that only a fraction of the information that made it through a threshold sensory memory of haptic, echoic, and iconic data would be consciously attended to and moved to a rehearsal space between working memory and durable brain storage. These cognitive models supported the managerial goal of HCI to ensure that the mind and computer are put to work in the most efficient way possible. For example, the design of computer interfaces using visual metaphors is intended to rapidly relate the computer worker to the work of the computer by appealing to (1) the worker's mental model and (2) the mental model designed into the operating system (both 1 and 2 need to be perfectly matched so that there is no confusion about what is being represented). The often cited and universal trash can icon or recycle bin metaphor on most computer screens demonstrates how a worker required to take part in waste management can do so intuitively and quickly by moving information from a storage area to a location ready for it to be deleted.

It is nonetheless significant that this metaphorical rendering of cognitive labor processes misses much of what actually constitutes work in the digital factory. The efficiency of the body is not simply replaced in the second paradigm by the disciplines concerned with the efficiency of mind. Instead, the flows relating to the labor of perception, attention, and memory are intimately interconnected to the flows of physical work. The labor of attention, for example, requires abstract cognitive functions to be aligned with the physics of brain power and the physical embodiment of the worker in computing environments. Bodies need to be situated in these environments just as they do in the mechanized factory. To be sure, throughout the first two paradigms and continuing into the third, this situatedness increasingly requires the energy expenditures necessary for long periods of physical inactivity. To sustain focused attention directed at screens and the repetitive inputting of information mainly using a keyboard and a mouse, worker inertia becomes a requirement of efficiency management. Distractions that draw attention away from the screen also lead to inefficiencies. The concern of cognitive HCI is therefore, at this stage, as much to do with inattentiveness as it is with attentiveness. The problem of attention, as Jonathan Crary argues, becomes ever more interwoven with inattention, requiring us to consider distraction as part of a continuum with the management of technologies of attraction.[24] As workers become more and more situated in front of screens (and connected to networks), they are increasingly grasped by HCI experts as likely to be inattentive to the task at hand because of the vast amount of information overload they experience in digital culture. "If we know that people are distracted, often involuntarily," a prominent HCI expert asks, "how is it possible to get their attention again without allowing them to miss the window of opportunity?"[25]

New Pathologies

In addition to the health hazards of an enforced sedentary work life, computer work, like mechanized work, also produces a body that more readily oscillates to the rhythm of the workplace environment. Along with the exaggerated vibrations found on the industrial factory floor, which persist to some extent in the whirring of hard drives, the clicking and tapping of external devices, and the endless buzzing of the call center telephone, there are further threats to a worker's health from physical and psychic pathologies in the digital factory. On one hand, we find both the habituations of repetitive physical tasks, such as mouse clicks and keyboard taps, and, on the other, stress caused by the quickening rhythm and information overload of work life. Indeed, the memory of the trade is not merely a matter of cognitive function alone but involves the repetitive interactions of a muscular memory: a mechanical habit that has a direct toll on the computer worker's body and mind, as is evidenced in ongoing medical concerns with repetitive strain injury and cognitive overloads leading to emotional stress.[26]

The Cybernetic User

The nascent trajectories of cognitive HCI have generally struggled to explain aspects of digital labor outside of the computer–mind metaphor. Certainly, in conceptual terms, the metaphor fails to grasp the social aspects of computer work beyond those located in mental models and information processing. The computational approach, for example, does not break with the computer–mind metaphor at all but rather transforms the coupled human–information processor into an entire system for managing efficient input. The so-called connectionists reject the specificity of the original metaphor, moving instead toward a new image of the brain as a neural network. Like this, cognitive subjects and computing machines become interconnected parallel nodes brought together to activate one another in a process of dispersed information flows. Similarly, the distributed model views cognition not as being locked in one person's mental state but instead as distributed across a number of agents consisting of human actors and computer systems. In other words, the social in both connectionist and distributed models merely becomes part of an information system through which cognitive mental models and information are circulated and differentiating representational states are shared and transformed, across a homogeneous system of coordinated components.

It is perhaps not until the pragmatic focus of usability studies that we see how cognitive HCI begins gradually to unravel as notions of cognition and information are jettisoned to be replaced by more socially (and emotionally) situated experiences. Although still generally focused on the mental requirements of individual users, emotions like frustration begin to work their way into the design of the digital workplace. The computer user is no longer, it would seem, considered a passive receiver of information. The user's interactions with the system, and others using it, become part of a more extensive working environment in which tasks are supposed to be carried out autonomously. Indeed, in sharp contrast to the inert cog in the machine of the ergonomic paradigm, the rhetoric of user-centered approaches within cognitive HCI tends to stress the autonomy of human agents in these environments. The agent has the capacity, it is claimed, to "regulate and coordinate his or her behavior, rather than being a simple passive element in a human-machine system."[27]

Taylor's Return, or Why He Never Really Went Away

The claims made by the usability movement within cognitive HCI concerning a worker's capacity to act autonomously are not, unfortunately, concomitant with experiences in the digital factory environment, which is now directly linked to consumption. Although it may be the case that buying online using a tablet device, for example, has become a less frustrating user experience—the disintermediated purchase decision usually only a few clicks away—the seamless connection between the consumer, the brand, the product, and the purchase masks a gruesome circuitry of invisible factories and workers: the digital equivalent of William Blake's *Dark Satanic Mills*. Even in the developed world, the reality of digital labor is far removed from the rhetoric of usability concerns. The widely reported working conditions experienced by "pickers" in the online retailer Amazon's storage warehouses, for instance, point to stark continuities between the digital factory worker and Gramsci's trained gorilla. The Amazon worker collects customer orders on a trolley using a handset scanner, which allots him just a set number of seconds to find each product. This is a coupling of bodies, minds and information that requires efficient movement through real and virtual workspaces simultaneously. Workers have described their experiences with these devices as like being treated as a machine or robot:

We don't think for ourselves.... We plug our scanner in, we're holding it, but we might as well be plugging it into ourselves.[28]

The brutal efficiency management of the Taylorist workplace is, it would seem, alive and well in the post-Taylor cybernetic factory, with the same increased risks of work-related mental and physical pathologies.

The Digital Gateway to Experience

Despite these bleak continuities with Taylorism, discontinuities begin to appear as computer work and consumption merge in a more generalized digital factory model in which the interests of HCI efficiency analysis and interactive design begin to overlap with the remit of experience marketing. To be sure, access to work communication systems, personal services like banking and shopping, and contact with work colleagues, friends, and family via social media begin to merge into one or two mobile devices. In the twenty-first century, these devices become the digital gateway through which worker and consumer engagements are increasingly lived through and managed, leading some HCI researchers to argue that new approaches to understanding HCI need to be developed that examine the *felt experience* of technology; that is to say, researchers should try to "interpret the relationship between people and technology in terms of the felt life and the felt or emotional quality of action and interaction."[29] The problem is that HCI research into felt experiences with technology is in perfect harmony with the marketer's evergreen pursuit of subliminal consumption. Indeed, in these latter days of the cognitive paradigm, there is an increasing interest in the unconscious processing of the emotional experiences of computer work progressively informed by new insights from HCI and the neurosciences.

THE EXPERIENCE PARADIGM

Pervasive Computing and Efficient Feelings

The arrival of a third paradigm of HCI is marked by two major trends. First, as a result of the continued miniaturization of computing components and mobility of wireless technologies, new pervasive contexts and ambient interfaces have been introduced to the digital factory. The most prominent technologies in this trend are radio frequency identification (RFID), the Global Positioning System (GPS), and a range of sensor and recognition technologies. Second, a previously marginalized set of social factors concerning emotions, feelings, and affect now figures writ large in HCI research. Both of these trends relate to novel embodiments and situated experiences of computer technology "whose central metaphor is interaction as phenomenologically situated"[30] and that are increasingly available in the workplace, the classroom, and the home.

In the first instance, the adoption of ubiquitous and pervasive computing technologies, like those using sensors and recognition software to detect active and passive interactions, brings about a new dynamic use context, aka the *Internet of things*. These new ubiquitous contexts were previously the subject of task-based heuristic user testing carried out in usability labs but now require a research focus on the cultures of usage of the kind usually carried out in ethnographic field studies. As computing continues to pervade these new social contexts, it becomes, as the interaction designer Adam Greenfield describes it, a process that is *everyware*, through which the intimate details of our lives are traded in for the convenience and accessibility of ubiquitous human–computer interactions.[31] Therefore, it is the goal of the digital business enterprise to, on one hand, provide more and more seamless interactions in the workplace through wearable RFID- and GPS-enabled computing linked to databases used by pickers in major supermarkets[32] and, on the other hand, ensure that pervasive computing reaches out to all consumer locations, including the shopping mall and the home.

It is difficult to apply conventional usability studies to these ambient interfaces because, on one hand, they are not necessarily oriented toward specific task-based interactions. In many cases, this is precisely what the system is devised to avoid—needing instead to respond to nontask interactions. On the other hand, people using ubiquitous computing are not required to contemplate, pay attention to, and view interfaces in the same way they did with conventional computing. The third paradigm claims, as such, to provide fresh insights into the way we think about interfaces. As Harrison et al. contend, the idea of embodied interaction moves on from second paradigm notions that grasp "thinking [as] cognitive, abstract, and information-based to one where thinking is also achieved through doing things in the world"[33] (the conventional notion of interface design based on seeing, hearing, and motor control of our hands is, like this, being reconsidered in light of technologies that support other senses and gestural manipulations). In fact, ubiquitous computing experiences are not necessarily supposed to have visible interfaces. They can readily dissolve into the surrounding objects and environments in which work and consumption are experienced.

In the second instance, the convergence between marketing, interaction design, and ubiquitous computing needs to be grasped in conjunction with new research interests in the changing social contexts of computing technology, primarily informed by an emotional turn in the neurosciences and digital workplace studies, wherein the focus of HCI is "pushed beyond limited domains of application and typical notions of 'work.'"[34] Initially, emotional, affective, and felt experiences were grasped as outside the remit of HCI because they could not easily be assimilated into the cognitive coupling of the human–information processor model. Questions concerning how people feel about interaction, the context of interaction, and elusive aspects of everyday life such as "what is fun?" were left at the margins of much of cognitive HCI research.[35] However, by drawing on a heady brew of phenomenological and neuroscientific interventions into what were deemed to be a naive Cartesian dualism at the core of the computer–mind metaphors of the cognitive paradigm, some HCI researchers have grasped emotional embodiment as a property of interaction outside the information coupling model.[36]

This contra-Cartesian trajectory in HCI research has been greatly influenced by Antonio Damasio's work from the mid-1990s, which positions Descartes as the straw man of an emotional paradigm in brain science. Damasio's thesis argues, in short, that emotions and feelings may not be intruders in the bastion of reason at all; they may be enmeshed in its networks.[37] Reasoning and decision-making processes are therefore not as purely cognitive as the second paradigm imagined them to be.

What Makes Them Click?

The significance of what appears to be a porous relation established between emotional and cognitive processing has been widely seized upon by business enterprises looking to steer decision making and purchase intentions relating to software and other everyday commodities. Much of this inspiration has been triggered by design gurus. Most notably, Damasio's thesis has been adapted for UX design by Don Norman, whose book *Emotional Design* argues that by factoring in user emotions, designers and marketers can capture consumer loyalty and engagement.[38] Norman points to the significant influence visceral, affective encounters with commodities can have on the processing of cognitive reflections and behavioral use. Not only do attractive things seem to work better, he contends, but a designer's appeal to emotions can bring the consumer and the brand closer together. The rise of a global UX industry can be grasped, as such, as a point where all the components of the third paradigm of HCI, including embodied interaction, felt experiences, emotion and affect, and a keen interest in the potential of neuroscientific market research, intersect with the cultural circuits of capitalism.[39] This is an intersection that becomes highly visible at the numerous industry and academic workshops that position UX as a central component of business success. UX design gurus like Norman are joined by UX consultants like Susan Weinschenk, aka the Brain Lady, whose book *Neuro Web Design: What Makes Them Click?* introduces a crude but resolutely business-friendly triangulation of the emotional brain thesis. This brings together the labor of online consumption, the unconscious processing of the old brain (mainly via the amygdalae), and a marketing-oriented mode of interaction design:

> There is an entire branch of marketing now focusing on activating the old brain and then feeding it product information.... Since a major job of the old brain is to keep us from harm, anything threatening our survival will get the old brain's attention.[40]

One answer to the question of what makes them click is, according to the Brain Lady, "*do something threatening.*" Significantly, though, fear is not the only emotion that can be activated to grab a computer user's attention and make her click. According to Weinschenk, access to the old brain, and subsequent admission to self-reflection, behavior, and purchase intent, can be tapped into via a gamut of emotions. A more general viscerality of felt experience associated with attractiveness, sociality, hunger, sex, and having fun can be employed to aid the experiential design of products and brands and therefore develop more intimate relations with consumers. UX design does not, for example, approach the design of an app from a functional point of view alone but rather designs for the emotional experience the application is supposed to trigger. The goal is to dissolve the product—the casing of the smart phone and the content and technical functions—into a desirable felt experience. Increasingly, access to this visceral level of experience processing is achieved by way of a range of user-centered empirical research methods intended to bring the emotional brain into the design and production cycle. These methods include ethnographic-like studies intended to tap into a user's emotional journey that can be correlated with biometric data garnered from eye tracking, galvanic skin response, and EEG. Like this, the third-paradigm focus on a computer user's felt experiences draws him into a production process that surpasses the reach of the cybernetic information flows of the second paradigm, extending outward to a far more exploitative and supple occupation of the entire sensory environment in which software is consumed. This is the emergence of neurolabor and consumption, the managerial focus of which is on efficient feelings and that is understood through the exploitation of neurological research into the diseases of the emotional brain, including attention deficit, obsessive compulsion, and addiction.[41]

The Ersatz Experience: Work Is a Theater, Business Is a Stage

Two significantly intertwined components of the experience paradigm need to be theorized in a political context. The first concerns the way in which capitalism is endeavoring to bring into play neuroscientific ideas to tap into the sensory environments of workers and consumers. Here we see the production of *mood environments* in which desirable, emotional, affective, and feely experiences can be captured and recycled so as to condition future user performance. The second revisits Nigel Thrift's use of the term *technological unconscious*[42] to describe a more neurologically oriented digital environment that combines with the experience paradigm to ensure that the conditioning of user performance is more rhythmically attuned to the needs of the market—made, as such, more seamless and, therefore, more efficient.

It is possible to comprehend how the first component encourages efficient future performances of digital workers and consumers by way of a resuscitated Tardean theory concerning the relation between desire, social invention, and imitation.[43] That is to say, there is a capture of the desires of workers and consumers associated with affective felt experiences that are subsequently appropriated by social invention and passed on (or recycled) as imitated ersatz experiences. In other words, the reinvention of experience becomes endemic to an artificial sensory environment, affective atmosphere, or imitated world in which visceral experiences are rehashed, and sold on, so as to trigger reflective thoughts and behavioral actions favorable to more efficient performances. A Tardean reworking of the experience economy can indeed be grasped, like this, as a theater of imitation in which sleepwalking performers are directed across a stage by a mesmerizing dream of action that leads them to believe that their desires and volitions are their own. To be sure, once the stage is invested with enough emotion, feeling, and affect, the performance itself escapes cognitive governance, only to be guided by a collective mood and thus managed by an *action-at-a-distance*. This production of collective moods opens up the potential for further "plumbing [of] the non-cognitive realm" of performance[44] and subsequently boosts the enthusiasms and motivations necessary to encourage future performances. This is a production line born not out of the ideas of Henry Ford but through the business of selling experiences in the same way that the experience economy pioneer Walt Disney did through the concept of the amusement park.[45]

The second component develops a previous notion of an expanding technological unconscious, which is dovetailed here with a somnambulist theory of HCI.[46] Indeed, I want to extend these ideas to more fully encompass noncognitive modes of HCI as they appear in the more advanced circuits of capitalism as forms of ubiquitous computing, specifically, a coupling between human bodies (brains, ears, eyes, thumbs, etc.) and all pervasive digital machines that manages not only to route around cognitive interactions but also to alter radically the relation a computer user establishes with her tools, typically characterized by the ergonomic paradigm. Pervasive computing is a reengineering of the human–machine coupling allowing for a much smoother, more passive, and potentially more rapid turnover of performances compared to the clunky machines of the ergonomic past. Digital ubiquity will undoubtedly help the business enterprise permeate nearly all aspects of everyday life as new modes of interactivity surreptitiously invade previously untouched social spaces. This is because human interaction with digital technology no longer requires a direct encounter with hardware or software but is experienced by way of previously unfamiliar situated experiences that further blur the distinction between production and consumption already conceived of in business literature in models of co-creation and customer made. Like this, the staging of user performances through pervasive computing potentially exacerbates the blurring effect to a point where the worker–consumer distinction dissolves altogether.

Ubiquitous computing also intensifies the managerial efficiency drives of the second paradigm by taking the principles of Web analytics outside of the networked PC environment and distributing them among the embedded objects and surfaces of *everyware*. Ersatz experiences can be filtered through ambient informatics and computational awareness (information about us) and produced by pervasive data gathering and location-aware technologies, including recognition software working on movement, sound, faces, or body heat. The question of what makes them click, once assumed by online marketing analysts to exist in the correlation between cursor movement and user attention, can now be answered by way of a coupling of neurological data about experience and the spatiotemporal location-based detection of moving bodies. This coupling presents a considerably deeper manifestation of the technological unconscious in terms of neurological mapping and the geographic location of users. Indeed, the question of user agency addressed in earlier incarnations of HCI is not so much concerned now with clicks as it is with the whereabouts and general state of the brain–somatic relation. In other words, the online marketer no longer needs to follow the intentional mouse click or a keyboard tap but instead focuses on the often spontaneous movements and emotional states detected when a person has entered (or not entered) a particular zone of interaction or leaves behind him an assemblage of personal data. Similarly, the data mining of online transactions, fed into databases and extracted as patterns for prediction and future suggestion, is superseded by systems that prompt movement in real time, speeding up the time spent between predictions, suggestions, and fast visceral thinking and action (the purchase). This unfolding of the technological unconscious sets the rhythm of work and consumption, capturing along the way the kind of brain time Gramsci once considered necessary for expressions of nonconformity.

The Gamification of Everything

The fusing together of ubiquitous computing and affective conditioning can be readily observed in the concepts and practices of gamification. A steady flow of business-focused literature has explored the idea that game mechanics can (1) create an upsurge in consumer engagement, (2) introduce behavioral change, and (3) increase productivity in the workplace. Indeed, as the technological wherewithal of pervasive computing proliferates into the sensory environment, we might expect to see the gamification of everything, or at least the wide-scale introduction of games to nongaming contexts in work and shopping situations, for example. This is a trend that the video game designer, Disney imagineer, and advocate of gamification Jesse Schell predicts will persist:

> We're, before too long, going to get to the point where every soda can, every cereal box is going to have a CPU, a screen and a camera on board it, and a Wi-Fi connector so that it can be connected to the Internet. And what will that world be like? . . . You'll get up in the morning to brush your teeth and the toothbrush can sense that you're brushing your teeth. So hey, good job for you, 10 points for brushing your teeth. And it can measure how long, and you're supposed to brush your teeth for 3 minutes. You did! Good job! . . . So you get a bonus for that. And hey, you brushed your teeth every day this week, another bonus! And who cares? The toothpaste company. . . . The more you brush, the more toothpaste you use. They have a vested financial interest.[47]

Schell goes on to imagine a future where a cornflakes packet with a Wi-Fi- and Facebook-connected Web game rewards you for eating your breakfast. There are bonus points, too, for taking the bus, where you can play a "REM-tertainment system" that "starts putting little advertisements out there to try and influence your dreams." Advergames have evolved into digital tattoos that work like Google AdSense but now use sensors to synchronize with other "tattoogles." There is also a new Kindle 3.0, which has the "eye-tracking sensor in it that can tell what you've read and how much you've read of the book." All of these seamless interactions will attract more and more rewards, while simultaneously steering attention to the ads and shopping baskets, clearly triggering a number of ethical problems for the designers of gamified systems. Nonetheless, despite these concerns about the level of transactional surveillance and clandestine management of attention, this stuff is, as Schell contends, *inevitable*:

> But these sensors that we're going to have on us and all around us and everywhere are going to be tracking, watching what we're doing forever. Our grandchildren will know every book that we read. That legacy will be there, will be remembered. And you get to thinking about how, wow, is it possible maybe that—since all this stuff is being watched and measured and judged, that maybe I should change my behavior a little bit and be a little better than I would have been? So it could be that these systems are all crass commercialization and it's terrible. But it's possible that they will inspire us to be better people, if the game systems are designed right. Anyway, I'm not sure about all that, but I do know this stuff is coming. Man, it's got to come! What's going to stop it? And the only question I care about right now is who, in this room, is going to lead us to get there?[48]

Indeed, the ubiquity of gamified experiences presents something much more than just the colonization of everyday life by information technology. It is a mode of behavior conditioning that taps into the emotional experiences of gameplay (having fun, compulsiveness, addiction, etc.) and adds them to the familiar experiences of encountering commodities like Pop-Tarts and Dr. Pepper. There will be many attempts to reproduce these kinds of experiences, some of which will dip below consciousness, by way of ubiquitous computing and affective conditioning that are not as easy to discern as those experiences we have experienced through current modes of interactive media. This is a regime of HCI defined by the way it routes around cognition. As Greenfield argues, we may perceive these different forms of interaction as part of a homogenous and continuous paradigm, so seamlessly experienced that they will "abscond from awareness."[49]

The Battle for Attention

There is a struggle for attention going on. Free, unmediated brain time is in decline. No more staring out of the window on the train to work. The time between morning and evening TV has been occupied by smart phone entertainment. The thin slice of human attention is increasingly being grabbed and managed by market forces. No more daydreaming in the schoolroom either. The teacher has to contend with students anxiously staring into the screens of their smart phones, continuously checking their notifications on Facebook. Students feel compelled, it would seem, to keep in touch on social media for fear of missing out on something occurring in these ersatz worlds. Indeed, it would seem that the old school and university model is struggling to keep up with the experience economy of neurocapitalism.

What this fear of missing out evidences is the force of a kind of marketing power Stiegler identifies as neuropower, exerted through the experiences of attentive technologies that absorb the brain time of youths, transforming attention itself into *attention engines*.[50] Neuropower further introduces a decisive split between the objectives of critical thinking—to dare to think nonconformist thoughts—and the goals of marketing power, which encourage worker and consumerist conformity by way of, among many other techniques, appropriating desires by grabbing subconscious attention, steering it toward specified windows of opportunity, and triggering restless competition. This is a dystopian mode of marketing power more akin to Huxley's reworking of a Disney-like transcendental Fordism than to the old biopower model applied to factories and schools. Instead of work life beginning with the industrial instruction Henry Ford's early apprenticeship schemes guaranteed, it now commences with earlier encounters with the emotionally charged conditioning of ersatz experiences, increasingly channeled through the sensory environments of mobile media, gamified reward circuits, wearable tech, and neurologically managed attention spans. The only way the school and university are going to keep up with the market is by Disneyfying the entire student experience! These new feely encounters are the point at which the market further pervades the freedom of Gramsci's brain to wander the factory floor thinking critically. As Stiegler points out, in this battle for attention, people lose the capacity for critical thinking and a certain resilience to market forces nurtured through intergenerational systems of care.[51] Indeed, every second of brain time, even that residual freedom Gramsci found in the Fordist factory, is, it seems, soaked up by a process of conforming to market forces.

Instead of continuing to lecture on critical thinking, it would perhaps be more productive to follow Huxley to a much more entertaining and emotional experience created by the Professor of Feelies, Helmholtz Watson.

3

Control and Dystopia

As an exemplar of what we might tentatively call dystopian media theory, Gilles Deleuze's short essay on control has a horrifying story to tell, that is to say, the "terrifying news" that corporate power has a *soul*![1] This soul is, evidently, the corporate marketer, who has—since the publication of the post-Foucauldian "Postscript on the Societies of Control"—seemingly moved into every corner of life under capitalism. Indeed, some twenty-four years after its publication, we are perhaps better placed now to gauge the full extent of Deleuze's dystopia. By doing so, I'm sorry to say that I have even more terrifying news. As well as a soul, the corporation has gained a brain. Like the scarecrow from *The Wizard of Oz,* the corporate marketer finally got what he wished for, in the shape of the practices of neuromarketing.

In this chapter, I want to persist with a dystopian media theory approach intended to unpick both the discursive formations and prediscursive forces that underpin two particular manifestations of neurocapitalism. The discussion begins by mapping the claims of marketers onto the dystopian visions that inspired Deleuze's thesis on control, including the explicit influence of William Burroughs, but also acknowledging the implicit influence of Aldous Huxley. What I find in this diagram is a series of mostly irresolvable paradoxes appearing in the old Foucauldian power enclosures of the schoolroom and the workplace. These paradoxes are typical of dystopian visions that extend beyond Foucault's discursive power into new arenas of control that affect the neurochemical interaction, including, on one hand, neuromarketing technologies intended to stimulate the so-called *buying brain* in the shopping mall and online store but also increasingly encompassing big pharma neuropharmaceutical interventions, on the other. I argue that although these paradoxes make certain predictions of a dystopian future fairly unstable, they also help to propagate it, because the open-ended nature of the absurdity makes for a volatile territory and therefore renders control open to the creative forces of rupture. In short, control societies are arguably at their most controlling when deterritorialized, and control today is, as such, a kind of open-ended *neuropower* that thrives on its own contradictions.

Before setting out some of the main ideas that frame a dystopian media theory, we need to briefly distinguish it from other critiques of capitalist production. Unlike the grand dystopian abyss presented by the Frankfurt School, for example, which grasped capitalism through the ideological lenses of dialectical movements, what is proposed herein is a cultural pessimism located in a distinct movement of paradoxes, that is to say, understood through often unresolvable and contradictory mixtures of hard and soft control, opposition and acquiescence, slavery and freedom. The chapter approaches neurocapitalism through the dystopias of Burroughs's open control society, Huxley's transcendental Fordism, and more recent contributions by theorists like Bernard Stiegler and Jonathan Crary, who argue for a kind of neuropower Gabriel Tarde similarly hinted at in the late nineteenth century. Moreover, the discussion draws on various neurotechnologies used in marketing and big pharma that can potentially bring to life the joyful encounters Deleuze recognized in corporate marketing control and Huxley similarly described as a soma-induced *feely* world in which *everybody's happy now*. Indeed, by way of recent examples of brain wave technologies, psychostimulants that paradoxically increase attention and docility, and efforts made by social media marketers to make user emotions become contagious, I will argue that a kind of Brave New World is already upon us.

THE RISE OF NEUROMARKETING: THE SCARECROW GETS A BRAIN

Like much of the glut of popular journalistic portrayals of neuromarketing, an article published in the *New York Times* in 2008 titled "Is the Ad a Success? The Brain Waves Tell All" is in seemingly incontestable awe of the claims that unconscious consumption can be measured directly at the brain.[2] "Never mind brainstorms. These days, Madison Avenue is all about brain waves." In all fairness, the article does at first strike a note of caution by recognizing that "some consumer advocates question the role of biometrics in ad research." It partially acknowledges, as such, concerns over what it calls a "blending" of *Weird Science* and *Mad Men*, which "will give marketers an unfair advantage over consumers." But these concerns are summarily dismissed, not least because the passionate adoption of neuroscience by marketers is a logical response, the journalist claims, to the slowing of the U.S. economy after the banking collapse in 2008. Because neuromarketers are only really interested in helping the economy grow in these difficult financial times, there should be little cause for alarm. As one marketing representative puts it, neuromarketing does not aim to "meddle with normal, natural response mechanisms." It is simply interested in improving sales by way of a better understanding of what consumers are paying attention to, how they feel, and what they remember about a particular product. After all, as the business school gurus persistently remind us, we live in an attention economy in which the drivers of focused mental engagement are at a premium.[3]

The neurosciences are quick to defend their links to marketing. Robert E. Knight, the director of the Helen Wills Neuroscience Institute at the University of California (a chief science adviser to a Berkeley-based neuromarketing company), says, "We're not trying to predict an individual's thoughts and actions . . . [or] trying to input messages."[4] There are no electrodes fed directly into the brain. Participants are willingly rigged up to noninvasive brain-imaging technologies, galvanic skin response devices, and eye-tracking software. These tests are generally carried out in the very early stages of product research and development so that brands can be more readily primed for consumption. So in many ways, there is nothing new here. The goal of market research has not changed. Marketers have always sought to attract attention and affect emotion by conditioning and anticipating consumer experiences in advance with the intention of seducing and guiding decisions by mostly subconscious means.[5] The difference with neuromarketing is, however, that the subliminal data captured from these experiences purportedly come directly from the brain. As a result, the mechanisms that capture attention and steer emotions are not reliant on the production of spectacles assumed to trigger semiotic processes of meaning making in isolation. They also become more concretely correlated to cognitive and affective triggering processes intended to span the lifetime of the brand–consumer relationship. This is how the neuromarketer claims to be able get inside the buying brain and guide it, often subconsciously, toward purchase intent.

As uncritical and overhyped as it maybe, popular media discourse surrounding the emergence of a much broader neurocapitalism points to something very familiar about the kinds of control circuits and technologies of power consumer societies are subjected to in the twenty-first century. In a nutshell, there is a scientific intensification of marketing. To begin with, neurocapitalism is intrinsic to a recognizable trajectory of marketing power in the electronic age that has, in the past, sought to capture the attention and sway the emotions of the masses confined to their living rooms via broadcast media. However, today, the corporate soul has adapted to digital media technology so as to expand control as a kind of McLuhanesque central nervous system that pervades fragmented, yet always connected, social networks. Evidently, this kind of control does not need to be physically administered in the sovereign sense of violent coercion. It is furtively distributed to a generally enthusiastic population seamlessly and constantly wired into the circuitous flows of corporate media systems via networked mobile devices, transactional surveillance, and intelligence gathering.

In further contrast to the architectural enclosures of Foucault's disciplinary society, marketing control has become an open, oscillating force of entrainment that quantizes the rhythmic performances of attentive subjects. There has certainly been an intensification of the affective rhythms of everyday cultural attraction. We can grasp the added value the neuromarketer offers to the soul of the corporation as a potential expansion of this affective rhythm far beyond the reach of conventional mediated communication into new corporate worlds of user experiences gauged at the level of the brain wave. In comparison to the ad men of the mid-twentieth century, who, by and large, reached out directly to the masses via the television ad break, marketing power today has developed on the subliminality of product placement as a kind of indirect and Trojan-like virus that infects attention and emotions. Via mobile and ubiquitous network technologies, this marketing virus can travel further than interpersonal connectivity to drill down underneath the individual–mass pairing into infrapersonal affective communication flows that steer attentive processes and emotional consumer engagements toward specific goals. We might even see in this affective turn a trend toward the targeting of noncognitive consumption. Noncognitive control is defined here as a technology of power that aims to dip below or route around cognitive processes. It would indeed seem that the enemy of marketing power today is cognition.[6] This situation becomes ever more apparent in the many appeals made to emotional, feely and affective registers that are now assumed to inform rational, logical decision-making processes. Marketers tend to trade more in the things that motivate choice and change behaviors through appeals to glamour, friendship, career aspiration, sex, celebrity narratives, intoxicating sporting glories, fun and games, and idiocy. They look to spread pleasurable affective contagions in addition to fear and insecurity. These joyful encounters spread well on social media, guiding attention and provoking decisions in ways that look to evade purely cognitive modes of choice.

Oh the joys of marketing, Deleuze bemoans.[7] Today it taps into affective transmissions emitted by what we might call an enhanced networked *dividual,* that is to say, not just the *data banks* of Deleuze's societies of control but also the endlessly divisible brain–somatic chemistries of networks of neurons, neurotransmitters, molecules, atoms, brain waves: *the infinitely small society of monads.* Premediated persuasion, anticipatory reward systems, and brain absorption become the watchwords of an increasingly pervasive digital-neurological marketing power. It is, as a consequence, imperative that those who enter into this cerebrally reinvigorated soul of the corporation do so with a clear sense of (1) how the near future might be mapped onto the dystopian imaginations that inspired the control society thesis and (2) how a dystopian theory of media might address what Fuller and Goffey call a "broader conception of mediality" working beyond representation at "scales that are beneath the level of the whole body, working on brains, neural entrainment and physiologically potent chemicals."[8] Grasped in this way, neurocapitalism can be seen as an open circuitry of control: a point of exchange where the external stimuli of digital marketing and neuropharmaceutical interventions intersect with bodies and networks of living neurons. Indeed, this is a novel mediality that is no longer founded on electronic technologies alone. We also need to consider chemical technologies that facilitate what Fuller and Goffey describe as neuronal transmission systems, meshing with the loops and hits of online connectivity, catalysing circulation through the topologies of networks linking synapses, minds, emotions, techno-science, geopolitics, creating grey media for grey matter and vice versa.[9]

DYSTOPIAN MEDIA THEORY

The problem facing theorists who wish to critique this neurochemical enhancement of the corporate soul is that it will not inevitably plunge society into absolute gloom. Or at least, it will not seem like a repressive system of control society needs to fight its way out of—again, such are the joys of marketing. This is in part because neurocapitalism introduces technologies of power that can often tap into feelings and attentive processes *noninvasively*. On the surface, this is no Orwellian dystopia of violent mind control. There is certainly no need for Cold War–style electrodes to be directly fed into the brain because the soft controls of the dystopian imagination of Burroughs and Huxley have provided corporate power with a neuroscientific universe in which to live. Similarly, neurochemical interventions into attention processes and emotions are not necessarily forcibly administered by an external power. As we will see, the chemical control of inattentive children is, for example, generally requested by parents weary of bad behavior or self-administered by students seeking a cognitive advantage. Control today is difficult to locate. *There are no electrodes or truth serums.* The chances are that, like Huxley's *Brave New World,* most of the population will be happy to go along with the contradictory mixtures of hard and soft control, opposition and acquiescence, slavery and freedom: *the most successful dystopias are, after all, always dressed up as utopias.* Like this, the dystopic soul of corporate marketing and big pharma seems to become a series of self-serving absurdities. However, before dealing with these contradictions, dystopian media theory needs to be grasped from the outset according to its own internal contradictions.

We can begin to trace the limitations and potential of these absurdities through the observations Raymond Williams makes in 1958 of Orwell's occupation of a paradoxical position in dystopian literature in terms of offering a socialist utopia[10]—a situation that is similarly inhabited by other writers, including Huxley, Burroughs, and even Deleuze, albeit in different political-cultural contexts. To begin with, through Williams, we can define an Orwellian dystopian media theory as the humane communication of an extremely inhumane terror. In this sense, it is an appeal to a kind of decency but actualized in absolute squalor. It is, in short, a dystopian cry for a utopian alternative. Dystopian theory is, as such, generally an expression of utopian ideals (socialist in Orwell and decentralized anarchy in Huxley's mature work) but remains resolutely apprehensive of the repressive extremes of these utopian principles, which can self-evidently manifest themselves into a line of flight as deadly as any fascist horror. So dystopian media theory can be paradoxically in favor of such things as social equality and a denier of class categorizations but presents these social groupings as rampant and mostly inescapable. Huxley's dystopia is indeed a double negative utopia. It is resolutely anti-American capitalism (in part inspired by his disgust at Fordlândia[11]) but also taps into contemporary bourgeois anxiety concerning Soviet communism. Similarly, Burroughs was a frontier liberal who ironically (and prophetically perhaps) advocated an alliance between American capitalism and Red China as a way to resolve political conflict.[12] So perhaps dystopian media theory should be located in a post-Marxist world? But there are further contradictions to be noted. As Williams contends, some Marxists regarded Orwell as the writer of horror comics which he readily sold on to his capitalist publishers for fame and fortune.[13] To be sure, most dystopian writers and theorists hail from petty bourgeois backgrounds. They are writers born into privilege taking the side of their proletarian victim. Through the creation of aesthetic figures like Winston Smith, John the Savage, and Annexia, the dystopian writer becomes the voice of a victim, opening up a universe of crushed despair, but nonetheless the authors are mostly victims without direct experience of this despair.[14]

There should not, however, be any theoretical limitation placed on dystopian media theory because of its encounter with the paradox. We will surely miss the force of dystopia if we look to resolve its contradictions by way of negation. Again, unlike the Frankfurt School, dystopian media theory does not think dialectically. It is, as such, distinct from the dystopian communication theories set out by Adorno and Horkheimer, for example. Following Nietzsche instead, perhaps the forefather of the dystopic encounter, William E. Connolly, similarly argues that:

thinking itself becomes blunted and dull if it always tries to resolve paradoxes rather than to open up spaces within them and negotiate considered responses to them.[15]

DYSTOPIA AS ANALYTICAL TOOL

To begin to think through the relation between marketing control, big pharma, and the brain, it would seem necessary to briefly revisit the primary dystopian inspiration behind Deleuze's control society thesis. Indeed, such an appeal to dystopian fiction (and the *interfering* aesthetic figures it brings to life) becomes a useful analytical tool, because Burroughs's *Naked Lunch,* for example, proves to be a precursor to a different kind of control emerging in consumer societies to that found in Foucault's disciplinary societies. In short, the hard stone and steel architectures of the prison, schools, and factories of industrial capitalism give way to a seemingly softer corporate capitalism. This is also no longer merely a capitalism of production; it is a capitalism based on the soft sell marketing of products and brands. So what significantly materializes in *Naked Lunch* and in Burroughs's later essay *The Limits of Control* is a far more open circuitry of power that benefits from recognizing its own boundaries.[16] However, nearly three decades before Burroughs established his vision of open control, Huxley's neuropharmaceutical soma became an aesthetic figure through which a similar concept of soft control would also come to life, albeit via a seemingly pharmaceutical sensation. The soft power evoked by soma was already apparent in Huxley's dystopia prior to the drug itself being invented. You do not control a population by hitting them, he contended. "Government's an affair of sitting, not hitting. You rule with the brains and the buttocks, never with the fists."[17] Better to have a population on their buttocks than on their knees! You need not make people fight for you either. You need forced conscription to do that.

Translated to the societies of control we experience today, it is much better perhaps to entice people into the shopping mall than it is to make them enter the ideological battlefield. Indeed, people who shop often feel compelled to work at it without being told to do so. In fact, in contrast to many of the modern-day addictions to junk (drugs), for example, which come under the kind of scrutiny of authoritarian state control Burroughs despised, to be a shopaholic is positively encouraged: a positive use of a dopamine rush. Today a theory of control should perhaps be more alert to the influence marketing has on subsequent obsessive–compulsive disorders and addictions than are the workings of the ideological state apparatus. Compulsive shoppers do not feel that their addiction to consumption serves the needs of corporate capitalism, just their own sometimes guilt-laden desire for a joyful shopping experience (retail therapy) and a pleasant, anxiety-reducing chemical release.

But this illusory sense of freedom and joy was never enough for the controllers of Huxley's dystopian universe. The compulsion (and the chemical release) had to be unquestionably guaranteed. So it was that the government of *Brave New World* subsidized thousands of pharmacologists and biochemists to come up with the perfect drug: "euphoric, narcotic, pleasantly hallucinant."[18] Forget religious opiates. Soma had it all: "the advantages of Christianity and alcohol; none of their defects."[19] As a kind of protoneuropower, it might be said that soma achieved a distancing function greater than any Cartesian dualism could ever pull off. Descartes managed to separate the rational mind from the unruly body, but soma built an impenetrable wall between the actual universe and the virtuality of the mind. This was a clever trick of neuropharmaceutical hypnosis of which Cajal's Mirahonda would have been proud: making a disconnected but altogether happy mind–body relation take on the appearance of thinking rationally. Add to that the self-propagating addiction of soma—"Take it," insisted Henry Foster, "take it"[20]—and we see the true force of Huxley's concept of neuropharmaceutical control.

It was Burroughs, nonetheless, who realized that control has to be paradoxically incomplete.[21] It is pointless implanting electrodes into the brains of school kids, consumers, or workers. They would become meager machines that could only be switched on and off. Beyond that, there would be nothing left to control. Complete control is no control at all. As Burroughs argued, it produces tape recorders, and you do not control a tape recorder—you merely use it! So control needs to be neither soft nor hard but a mixture of the two. It needs to be left open: not closed, not whole, but an always divisible composite. Noninvasive control needs to allow for opposition and acquiescence to be able to break out together from disciplinary enclosures. Indeed, the terms of engagement for partial control were well understood by Burroughs. School kids, shoppers, and workers (all subjects of power) must, in contrast to machines, be treated like dogs under domestic hypnosis. They must be allowed off the lead, able to bark once in a while, and encouraged to explore outside their enclosure so that they can one day dream of escaping it. It is this same openness of control that Deleuze writes about in "Postscript on the Societies of Control." Fitted with an invisible collar and chain, the subject of power becomes a divisible composite in an open system of controlled labor. Just as care in the community releases patients from the hospital enclosure into an open circuit of control, the subject in the shopping mall cannot properly discern between freedom and slavery.

To open up these novel spaces of control in times of neurocapitalism, a dystopian media theory needs to follow a similar series of contradictions. First, there is a need to find the intersection where utopian virtuality overlaps with dystopian actuality. On one hand, capitalist utopia today is, as ever, built on the eternal promise of new and *inevitable* solutions and technological fixes to the economic woes of the latest financial crash and the *necessary* pain of austerity and indebtedness. Currently this eternal economic reinvention is driven, to some extent, by new attentive technologies (mobiles, tablets, apps, and other smart technologies) and bolstered by novel marketing strategies, including the emotional branding of the user experience measured by eye tracking and brain wave activation. Some business gurus call it the attention economy, others the experience economy. A dystopia is nonetheless, on the other hand, apparent in a sharp rise in the side effects of attentive slavery and emotional capture, including the well-publicized and seemingly toxic spread of mass attention deficit recently highlighted in Bernard Stiegler's account of the socially destructive force of marketing technologies.[22] There is indeed bleak statistical support for Stiegler's concerns.[23] The attention-deficit *epidemic* began in the United States with a reported 9 percent of children between the ages of three and seventeen diagnosed (4.7 million) in 2011. Similarly, in the United Kingdom, there was a 56 percent rise in the prescription of attention-deficit pharmaceuticals reported between 2007 and 2012. But herein a second contradiction lies, because although this increase is in part ascribed to adults as well as children being diagnosed with and treated for ADHD, it also corresponds with a rise in the use of psychostimulants as so-called *smart* or *cognitive-enhancing* drugs popular among competitive students seeking an intellectual advantage. So this pharmaceutical dystopia is not simply a poison. For some it is a cure for the kind of inattentiveness that jeopardizes exam results and a good career. Indeed, in the online neuropharmacy, we find the cure for the fallout from attentive slavery in the shape of smart drugs designed to treat attention disorders. Such drugs are actively prescribed and self-administered. Third, and relatedly, as attentive technologies become more ubiquitous, they help to produce more alert, productive, predictable, and efficient workers and consumers, but they also make schools more open to the distractions (and emotional diseases) of the market, requiring more diagnosis of (and cures for) attention deficit and media addiction.

If there was not a need then, as Deleuze contended in his analysis of control, there certainly is a need now, in times of neurocapitalism, to invoke neuromarketing technologies and the extraordinary pharmaceutical productions that increasingly target the brain, the neuron, and the dopamine rush, promising to enter the open circuits of control, to bring together the universes conjured up by dystopian fiction, in which liberating and enslaving forces confront one another in a single cosmos. Again, it is important to note that these paradoxical mixtures never become concrete. The buying brain, like the soul of the corporation, is always exposed to capricious chaotic forces. Control needs to capture and escape the potential of what passes it by. It is never fixed. It is at its most creative when deterritorialized. Openness exposes control to these irresolvable contradictions, making it simultaneously robust and vulnerable to breakdown and collapse. The promises of neuromarketing, for example, to be able to anticipate and guide the attention of docile consuming subjects are perhaps just that, it would seem: nothing more than promises made to shareholder investors. However, these are promises always backed up by invariable managerial efforts to tackle the evil of efficiency discussed in the previous chapter. To be sure, as the shadow of neuroculture passes over the cultural circuits of capitalism, which interweave the marketing seminar, digital workplace, shopping mall, classroom, and online drugstore, it becomes ever more rehearsed at producing contradictory, yet manageable, efficient, predictable, and divisible subjects who are, at the same time, emotionally distracted and attentive subjects. So although neuromarketing may well prove to be an imperfect marketing tool for grabbing attention and steering emotions, we still need to take seriously its claim to an objective psychology of consumption. Similarly, the neuropharmaceutical management of attentive and emotional states may prove to be a capricious form of control, but we still need to seriously consider its potential role in the production of docile subjects. Simply put, even if neurocapitalism only delivers a fraction of what it promises to deliver, then we need to be alert to a considerable deepening of Deleuze's societies of control. Indeed, before we ask what our brains can do, we need to ask what can be done to this objectified brain—*without electrodes.*

THE PROBLEM OF ATTENTION DEFICIT

Two contemporary theorists who alert us to the dystopian paradoxes of our current control societies are Stiegler and Crary. To begin with, and closely following Crary's thesis concerning how the capture of attention plays a key role in the logic of capitalism, we find that distraction needs to be seen as a constitutive element of the many attempts to produce attentiveness in human subjects.[24] Watching too much TV, playing video games to excess, or obsessively checking a smart phone all day becomes, in certain contexts like the schoolroom, a kind of distracted attentiveness. Primarily, then, modern distraction is not seen as a disruption of normative, stable, or natural kinds of attention. The problem of attention, as Crary contends, is inseparable from inattention. They are not polar opposites; they are a continuum.[25] At the center of this paradox, we find the increasing diagnosis of a dubious "disease" and the prescription of equally suspect neurochemical remedies. For Crary, attention-deficit disorders firmly position a certain kind of attention as a Foucauldian normative category of institutional power. But although the ADHD child cannot pay attention to his schoolwork, he can stare endlessly into the screen of his mobile, an essential tool for accessing the attention economy. Here the contradictory tensions between the schoolroom and the marketplace can be seen to intensify. Indeed, the efficient management of attentive subjects in the disciplinary enclosures of the school is, as Crary argues, "working imperfectly at best."[26]

Attention and inattention are certainly respected in different ways in education and marketing. For example, in the classroom, inattentive subjects generally become the anomaly, whereas in the attention economy, they becomes the norm by which consumers are measured. It is indeed assumed that consumers are inattentive and in need of focus.

In other words, certain expectations about how individuals should behave in disciplinary educational enclosures suggest that the inattentive are, to some extent, at odds with the necessary behaviors required to efficiently participate in the attention economy. Similarly, the collection of symptoms of ADHD ("impulsiveness, short attention spans, low frustration tolerance, distractibility, aggressiveness and varying degrees of hyperactivity"[27]) seem absurdly compatible with the banal mental performances, sedentary physical modes, and perceptual overloads of excessive screen-based attentive consumption, that is to say, a culture awash with distracting fixations, requiring short attention spans and technologically enforced docility.[28] Indeed, the relation established between attention deficit and paying attention to the market becomes a neuromarketing strategy. The neuromarketer has, as such, appropriated the pathologized inattention of the ADHD child. "It's kind of funny isn't it?" says one high-profile marketer. "A consumer in a store, or a consumer at a website, or a viewer sitting with a remote in their hand, represent to us the most attention challenged human being."[29] No surprise, then, that the neural markers of attention-deficit, obsessive–compulsive, and memory disorders, and the neurological tools used to measure them, become indispensable to the neuromarketer's tool bag.

It would initially seem that the paradox of attention and inattention is not so pronounced in Stiegler's approach to the problem of youth attention. There is still a growing tension between education and the marketplace that has, according to Stiegler, become part of a battle for attention, but there is a clearer distinction made between, on one side, education and care and, on the other, the market and carelessness. In fact, Stiegler tries to reposition the school as a site of struggle rather than a disciplinary enclosure. This is a struggle against the *capture, destruction,* and *reinvention* of attention by industrial populism.[30] In other words, the proliferation of attentive screen technologies (psychotechnologies) produced by the cultural industries is becoming a subversive force in society, wreaking havoc on the process by which a child learns through the formation of attention through care: a general education experienced in her focused relation to adults. Children, Stiegler contends, "deserve better than *that.*"[31] Similar to Crary, though, there is a paradoxical relation established between the capture of attention and the increase in cases of ADHD. It is absurdly the proliferation of attentive technologies that (1) diverts and attracts attention and (2) produces attention problems. This contradictory situation arises, Stiegler says, because a child only has a finite amount of brain time. Indeed, the demands on this cerebral time to consume result in a struggle first for attention and ultimately for intelligence. It is like this that the brain becomes central to the struggle for attention. The destruction of attention is directly linked to the manipulation of plastic synapses, that is to say, a "premature structuring and irreversible modelling of [a child's] synaptogenetic circuits."[32] The demands to consume in a market economy are thus grasped as a reengineering of the attentive brain, leading to the destruction of a child's "affective and intellectual capacities" in favor of the production of attention engines in the service of popular industrialism.[33] This is the making, Stiegler claims, of a new kind of proletariat.

THE TECHNOLOGIES OF NEUROPOWER

There are a number of further important crossovers between a dystopian media theory of control and what Stiegler calls *neuropower*. Speaking in Amsterdam at the Unlike Us conference in March 2013,[34] he expanded on this concept—a progression beyond Foucault's biopower and his own interpretation of psychopower to a power that circulates digital and cerebral automations. Significantly, neuropower is not reduced to the individual brain but is always in a *transindividual* relation to other brains, that is, a *bridge* that links the individuated brain to the collective. This should be no surprise, because although the neurosciences have a tendency to focus on individual brains, neurocapitalism is increasingly concerned with infrasubjective relations, such as those established in collective rhythmic entrainments. Like this, Stiegler's neuropower evokes the social ecology of brains that Gabriel Tarde's nineteenth-century media theory hinted at, that is to say, an ecology of mostly unconscious associative relational flows in which brains fascinate and polarize the desires and beliefs of other brains.[35] Tarde's early crowd theory focused mainly on social relations of imitation established between the biological cells in different brains but went on to note the early formation of mediated publics (newspapers and telegraphs), contrasting their docility to the magnetizing animal forces of the crowd.[36] Stiegler similarly extends neuropower to a physiological brain relation with synthetic social media networks, a kind of tertiary memory that goes beyond the primordially biological relation established between perception and memory toward an artificial and transformative relation established between brains and artificial networks. Indeed, Stiegler goes on to position social media as a surrogate artificial crowd. This is not a strictly biological brain in which memory is recollected as psychic or synaptic energizations (or brain traces) but is constituted as a digital recording of memory at a tertiary level. So neuropower comprises a convergence between, on one hand, a desire to intervene in neurochemical processes relating to the market, as expressed in neuromarketing and neuroeconomics, and, on the other, the artificial crowd composed of synaptic connections between humans and digital technology.

The examples introduced subsequently also correspond with Stiegler's neuropower insofar as biological, cultural, sociological, and technological distinctions are collapsed into a generalized concept of neuroculture. Indeed, the brain can be grasped, via Huxley's dystopia, as a site of mediation in relation to various neurotechnologies of power, specifically, EEG and neuropharmaceuticals like Ritalin and Adderall XR (a dexamphetamine competitor of Ritalin), all of which are, incidentally, used to *detect* and *treat* ADHD. Here we encounter controversies surrounding brain wave technology, intended to diagnose ADHD as well as correlate sensual stimuli with the behaviors, mental states, and rhythmic entrainments of consumer experiences, and, similarly, the potential over-prescription (and self-medication) of neuropharmaceuticals designed to target "defective" neurotransmissions (mainly dopamine receptors linked to attention and reward systems) to produce *normative* and accelerated attention. Finally, the artificial crowd, composed of synaptic connections between humans and digital technology, can be seen to facilitate a dystopic mode of control wherein attention is drawn toward content intended to trigger ersatz emotional experiences and contagions.

WEIRD SCIENCE: A BRIEF ARCHAEOLOGY OF THE BRAIN WAVE

In an effort to evaluate the extent to which his dystopian universe of control had become a reality in the 1960s, Huxley revisited the themes that had obsessed him in the early 1930s, including the potential manipulation of social organization through propaganda, brainwashing, hypnosis, and chemical persuasion.[37] The eventual realization of these technologies of power, he contended, was underpinned by one common denominator: Pavlovian conditioning. In fact, from the evidence he collected, Huxley saw Pavlov as integral to an "ultimate revolution" in control, extending from the violent constraints of terrorism to the scientific tyranny of what might appear to be a benevolent dictatorship. It is not that terroristic control would disappear altogether. As we know today, under the stress of post-9/11 terrorism, a population can be made more susceptible and absorbent to political conditioning and control. But it is important to note that stimulus conditioning, including the conditioning of emotions other than fear, is an anticipatory mode of control reliant, at first, on the analysis of a population through demographics and polling data. Indeed, Huxley's ultimate revolution begins by transforming the old populations of sovereign and disciplinary control into statistical populations, without the need for absolute violence or enclosure. Amid Cold War paranoia, research into susceptibility to brainwashing, hypnosis, and the placebo effect produced startling statistics that both fascinated and horrified Huxley. For example, and echoing Mirahonda's assessment, it was estimated that approximately 20 percent of any given population were vulnerable to suggestion. Huxley bemoaned that, in the hands of a dictatorship, this 20 percent could easily become a powerful weapon against democracy. Certainly he recognized the use of mass media networks to ensure that hypnotic control could spread more readily to the masses. But Huxley's control society required more than just mediated propaganda and market research. His Fordship, Mustapha Mond, the Central Controller of *Brave New World*, wanted to know what could be done to a brain.

A brief archaeology of early brain technology evidences the absolute theoretical futility of making any distinction whatsoever between so-called normal and weird science. *All science is weird science!* Indeed, in the period between 1932, when Huxley wrote his fable, and the late 1950s, when he revisited it, there had been a strange revolution in brain science and associated neurotechnologies. The neurosciences not only uncovered brain anatomy and functionality but also began to directly intervene into brain processes. Following Richard Caton's observation of brain waves in 1875 and Hans Berger's identification of certain frequencies in human brains in 1924, and the subsequent invention of EEG to record these waves, Huxley notes a sharp acceleration in invasive and noninvasive incursions into the pleasure and pain centers of the brain. Invasive and noninvasive EEG records the frequency of brain waves relating to the neuron-to-neuron flow of electrical impulses in brain regions during sleep, relaxation, and active modes.[38] Although in the Cold War–fueled scientific fervor of the mid-twentieth century, rats, chickens, and the insane were apparently subjected to Pavlovian remote control conditioning of physiological and psychological responses by way of electrodes fed directly into their brains,[39] Berger had already established noninvasive methods in the 1920s. He started out by inserting clay electrodes and steel needles directly into the brain but was soon able to record brain waves by attaching electrodes to the scalp using adhesive tape.

By the late 1930s and early 1940s, EEG researchers had identified four kinds of brain waves correlated with various conscious and unconscious behavioral states. To begin with, it would seem that neuron-to-neuron flows are never switched off. Delta waves are indeed present when a person is in a deep dreamless sleep, indicating that brain waves are at their slowest (1–4 Hz). Theta waves (4–8 Hz) show that a person is in a dreamy, light sleep, associated with REM sleep and trancelike states. Alpha waves (8–12 Hz) are present when a person is in a woken but relaxed state, linked to the production of serotonin, which helps relay signals across regions of the brain. Finally, beta is a fast brain wave frequency (12–30 Hz). It is related to attentive and alert states, including learning, but also anxiety, fear, and stress. Indeed, making a link between the practices of hypnopedia in *Brave New World* and early EEG experimentation at the time, Huxley points toward a zone of hypnotic susceptibility in between light sleep (theta) and relaxed woken states (alpha), in other words, when alpha brain waves start to become present, mixing together semiconscious trancelike states and lighter dreamy conscious states.[40] For example, when exposed to loud noises, a person in a deep, dreamless sleep (delta) will be suddenly awoken, exhibiting beta brain waves. In contrast, a less violent stimulus will not arouse him completely but will cause the reappearance of alpha waves mixed in with theta waves.

Another significant parallel between Huxley's hypnopedia and brain wave frequency research first begins to surface in EEG experiments in the 1950s and in the revelation that some alpha and beta waves can be synchronized to external rhythmic stimuli, including strobe lighting and rhythmic noises. In 1953, the neurosurgeon Gray Walter reported that under certain conditions, people would enter trancelike states in which they experienced serenity and dreamy visions.[41] Indeed, Huxley would have been very interested in more recent EEG research into brain wave entrainments triggered by rhythmic music. Although not fully understood, external stimuli of this kind can induce *frequency following* (neural assemblies characterized by the synchronous activities of constituent neurons).[42] The link between brain wave entrainment and affective states, although again not fully understood, seems to underpin the efficacy of music therapy intended to treat attention deficits by synchronizing brain frequency to external audio stimulation.[43] In *Brave New World,* Huxley similarly notes how the power of the "beating of drums and a choir of instruments—near-wind and super-string . . . plangently repeated and repeated" produces an inescapably haunting melody capable of gathering the attention of the Brave New Worlders and quelling any misguided thoughts of riotous assembly. These rhythmic musical entrainments are also linked by Huxley to affective states. He writes that it was not "the ear that heard the pulsing rhythm, it was the midriff, the wail and clang of those recurring harmonies haunted, not the mind, but the yearning bowels of compassion."

There are further lines of flight running between Berger's invention, Huxley's concerns in the 1950s, and the current problem of attention as it is approached by both neuromarketers and the neuropharmacy. On one hand, EEG has become the tool of choice for the neuromarketer.[44] Like this, it models the consumer in a store, on a website, or using mobile devices, as attention challenged. EEG is therefore used to make correlations between the frequency of brain wave signals and the cognitive and affective states subjects experience in the sensory environments of the marketplace. On the other hand, correlations made between inattentive and hyperactive behaviors and particular brain waves detected by EEG become endemic to more recent controversies surrounding the clinical diagnosis of ADHD. In the United States, an EEG, said to help identify attention deficit as a problem with brain frequency, was approved for marketing in 2013.[45] This technique in effect compares the electrical impulses generated in neuron-to-neuron communication with a child's capacity to pay attention in the classroom. In short, the ratio between theta and beta brain wave frequencies in particular regions of the brain is supposedly higher in children identified as suffering from ADHD than in those subjects tested in *neurotypical* control groups.

What these trends in diagnosis point to is, on one hand, a solipsistic assumption that attention deficit is a problem to be located *inside* the brain of both the consumer and the schoolchild, not, it must be stressed, in the sensory environments in which they consume and learn. As the neuroscientist Steven Rose argues, in the schoolroom, the "cause" is "unequivocally located inside the [ADHD] child's head. Something is wrong with his brain."[46] On the other hand, though, the measuring of brain wave synchronizations, according to both the excess flows of marketing messages and the "normalized" attention spans of the schoolroom, can be seen as consistent with the history of industrial working rhythms formulated by E. P. Thompson back in 1967.[47] Thus, we can map these frequency-following disciplines to a trajectory that perhaps begins in the sensory environments of preindustrial life (church time, bell ringing, sacred calendars, etc.)[48] to the onset of capitalism and the introduction of industrialized factory time, wherein clocks, machine cycles, twelve-hour shifts, and, eventually, the inception of the Taylorist time and motion workplace replaced the time of the natural rhythms of the seasons and daylight working hours of rural work. These latest attempts at brain wave entrainment can, as such, be seen as an extension of capitalistic disciplines of temporality and speed also comprehensively addressed by Paul Virilio.[49] Indeed, the increasing interest in the rhythmic frequencies of the consuming brain and the normative range of attention required for school and work are endemic, I contend, to a realization that the *speed factories* of capitalism require a well-synchronized and entrained workforce, originally organized around physical and psychic labor but now, increasingly, through the quickening of neurological energies attuned to sensory environments. The rhythmic entrainments of neuroculture become the latest territorializing force of capitalist subjectification—a unification and acceleration of workers and consumers Huxley well understood and expressed through the aesthetic figure of *The Second Solidarity Hymn*:

> Ford, we are twelve; oh, make us one,
> Like drops within the Social River,
> Oh, make us now together run.

THE FOUR PARADOXES OF THE NEUROPHARMACY

It is nevertheless the dramatically expanding universe and intervening powers of neuropharmaceuticals that troubled Huxley the most. In 1931, when he was writing *Brave New World*, the study of brain chemistry was in its infancy. Yet, by 1958, the "non-existent ripple" caused by the discovery of neurotransmitters, like serotonin, had turned into "a tidal wave of biochemical and psychopharmacological research."[50] Pharmacology, biochemistry, and neurology were by the 1960s well and truly on the march. New and improved pharmaceuticals, with reduced side effects, that could increase suggestibility and lower psychological resistance to control were a distinct possibility. To be sure, Huxley was convinced of the prospects of a control drug emerging in the 1960s that would combine the dulled happiness and hallucinatory visions of soma. He even goes on to develop on a future scenario of pharmacological control wherein an *impartial* science produces a drug that is initially intended for benevolent psychiatric care but eventually falls into the hands of a malevolent dictatorship. Such a drug would tap into the common desire to "take a holiday" from the black moods of depression ("a gramme is better than a damn") while also ensuring that the mantra of the Brave New Worlders ("everybody's happy now") remained loud and clear, even in the most abhorrent of circumstances. As Huxley argues, "the soma habit was not a private vice; it was a political institution." Like the cosmetic pharmacology of *Prozac Nation,* it would serve a paradoxical political purpose to both "enslave and make free, heal and at the same time destroy."

So, in times of neurocapitalism, how close do we come to Huxley's paradoxical psychotropic paradise? In other words, is it the case, as Rose argues, that the desire for a multitude of neuropharmaceuticals will usher in a Huxleyesque society determined by (1) state prescription of drugs that control behavior and (2) a consumer need for designer drugs that enhance cognition?[51] Certainly, in addition to the well-documented rise in Prozac prescriptions, the increase in the diagnosis, chemical treatment, and off-label usage of ADHD drugs like Ritalin and Adderall XR points to a potential manifestation of a dystopian future vigorously marketed by big pharmaceutical corporations and popular with students and technology start-ups craving attentive advantages in the schoolroom and the digital workplace. This is a dystopian future characterized by a further series of paradoxes.

First, the delineation of normative and anomalous attentive behavior, in the schoolroom and the marketplace, compares to Crary's continuum in which inattention becomes a constituent element of the many efforts made to produce attention. That is to say, anomalous behavior in the schoolroom—inattention, restlessness, impatience, and aggressiveness— can often spill over into the normative range of behaviors and mental activities necessary to participate in the attention economy. Indeed, given the close proximity of these behaviors to both unmanageable school kids and good consumers, it is perhaps "nonsensical to pathologize" ADHD, because, as Crary puts it, we live in a culture that is

> relentlessly founded on a short attention span . . . on perceptual overload, on the generalized ethic of "getting ahead," and on the celebration of aggressiveness.[52]

Indeed, the endeavor to locate attention deficit in the neurochemistry of the unmanageable schoolchild is regarded by Crary as the production of a convenient yet imaginary disorder.[53]

Second, and relatedly, the location and targeting of certain neuro- chemicals by ADHD drugs are justified by often conflicting expla- nations of how these chemical interactions actually function in the brain. Although attention deficit is commonly attributed to a fault in dopamine neurotransmission, brought about, in turn, by an assumed but as yet unknown genetic predisposition, dopamine itself seems to have many contradictory functions, including some that would be of interest to Huxley. To be sure, on one hand, the neurochemical is widely regarded to be engaged in the switching of attention and the selection of action or behavioral invigoration and management.[54] It is related to saliency, that is to say, a "rousing event to which attentional and/or behavioral resources are redirected."[55] On the other hand, though, and more relevant to Huxley's Pavlovian world of control, is dopamine's suspected association with behaviorist reward systems. As a result, it is often called the hedonistic chemical of pleasure and implicated in behaviors and decision-making processes linked to neurological re- search into addiction. When administered drugs that block the effects of dopamine, lab rats, for example, will tend not to bother pressing a lever for food. In contrast, shots of cocaine and amphetamine, which interact and synthesize dopamine neurotransmitters, will provoke the rat into choosing an advantageous place in which to wait for more drugs. Like this, Joseph LeDoux theorizes that dopamine's interactions with the amygdala, the hippocampus, prefrontal areas, and the motor cortex combine to produce emotional arousals that motivate action and decision making in humans.[56] Corresponding to Pavlov's stimulus and Skinner's instrumental conditioning to some extent, dopamine seems to exert an anticipatory rather than consummatory mode of chemical control. Significantly, the chemical is not released at the point when Pavlov feeds his dogs or the rat gets his next fix but in the moment when the bell rings for dinnertime or the lab rat moves to a location where he can advantageously wait to satisfy his addiction.

Third, ADHD drugs seem to exist somewhere on the same pharmaceutical cure–poison continuum that soma occupies. On one hand, regarded as a cure for ADHD, these drugs are assumed to modify undesirable behavior by simultaneously tranquilizing and boosting the attentive faculties of the ADHD child, presenting a paradoxical remedy for distraction. The clinical recognition that psychostimulants, including methylphenidate, dexmethylphenidate, and amphetamine-dextroamphetamine, can at the same time pacify and stimulate was indeed the main supporting hypothesis behind ADHD. The question is, why were these distracted and disruptive children absurdly affected in this way? Well, as Rose points out, more rigorous testing has established that there is no specific ADHD drug paradox. It was indeed a widely promoted assumption overturned when it was discovered that psychostimulants affected control groups in exactly the same way.[57] Notably, the paradoxical effects of stimulants, like Adderall XR, have been recorded in EEG testing, wherein alpha and beta frequencies increase in chorus with each other.[58] On the other hand, ADHD drugs have toxic side effects. Their impact on short-term behavior in the schoolroom is fairly predictable, but long-term effects on mental and physical health are less clear. The mental side effects of Adderall XR can include aggressiveness and paranoia, whereas physical side effects manifest as rashes and tics, a considerable loss of appetite, heart palpitations, and potentially fatal heart problems. Moreover, although many children on Ritalin report feeling calmer and more concentrated, and subsequently work harder at school, making life easier for harassed teachers and parents alike, some have reported that taking the drug outside of school makes them feel socially isolated.[59] Similarly, a common side effect of Adderall XR is depression, including sadness, hopelessness, insomnia, and a loss of interest in social activities.

Fourth, the realization of a soma-like neurochemical control has to be grasped alongside the unintended consequences of these side effects and subsequent *off-label* uses. Adderall XR was originally intended to treat childhood obesity. As a type of amphetamine (or speed), it reduces appetite, leading to weight loss. But like many pharmaceutical products in an overcrowded drug market, Adderall failed to catch on, until, that is, its side effects were identified as having a calming influence on hyperactive behavior. Adderall also triggers affective arousal, stimulating cognitive states related to alertness, wakefulness, and, of course, enhanced attention, but it was its calming effect that apparently drew the interest of the pharmaceutical company Shire AG, which bought the rights to the drug and repositioned it in the marketplace as a treatment for ADHD.[60] It is similarly claimed that Ritalin boosts neurotransmitter levels primarily in the prefrontal cortex, positively affecting cognitive processes, including attention and chemical reward mechanisms. However, there is still much uncertainty as to how Ritalin affects the brain chemistries it interacts with.[61]

The side effects of neuropharmaceuticals may remain purposely open to the "changing fashions on psychiatric diagnosis,"[62] but the off-label uses of psychostimulants help to propel them beyond the discursive formations of the neuropharmaceutical corporate giant into a desiring economy of self-medication and consumer choice. Made accessible through the growing number of unregulated online pharmacies or by way of illicit means, such as through students faking the symptoms of ADHD in the doctor's office, Adderall has become the preferred and most widely distributed "study buddy" or "academic steroid" in U.S. schoolrooms.[63] Similarly, another speed-based drug, Provigil— initially intended to treat certain circadian rhythm sleep disorders mainly caused by shift work patterns—has become an off-label cognitive stimulant used by workers, students, and soldiers alike. It is, as such, a popular nootropic drug well suited for the demands of the attention economy. It assists in meeting the requirement to stay alert, attentive, and memorize more and more sensory data and thus rewarding its users with an intellectual advantage: "*if everyone else is doing it why shouldn't*

I get the advantage?"[64] Nootropic drugs are, moreover, endemic to a control mechanism that evades a Huxleyesque Central Controller. To be sure, to control, it is pointless, as Burroughs realized, to make the population into machines that can be switched on and off. Opposition and acquiescence, freedom and slavery, must be retained in the control system. Complete control is no control at all. *The population needs to be able to turn itself on!*

WE'RE ALL GOING ON A SOMA HOLIDAY

Stiegler's work in fact rallies against the kind of drug-fueled scientific purification Cajal's Mirahonda encourages: seeing Ritalin as the hyper-proletarianization, or a chemical hypnosis of the worker, which keeps people happy in the factory without them having to think.[65] The pharmacology of the attention economy (the cure and the poison) is not therefore merely a symptom of the destruction of attention; it is also part of a social reinvention of an already rigid class system by way of a dulled sensation of conformity. Similarly, in *Brave New World,* soma is a drug that produces the necessary docile subjectivities that conserve a hideous social class system. Helmholtz Watson, in his augural lecture at the College of Emotional Engineering, would perhaps add a historical footnote to the origins of soma, outlining how the reappropriation of ADHD-type drugs begins in the elite university towns: sites of power where Mirahonda's good brains were an outcome of both good fortune and smart drugs. The adoption of these drugs by what would become the Alpha caste leads inevitably to more liberal attitudes. Well, no one wanted to be left behind in the rush to get smart. Governments would eventually learn that it is more popular, affordable, and convenient to deal in these drugs themselves rather than legislate against them. The Alphas were, after all, a population already self-medicating with morphia and cocaine, "creating a kind of hypnosis, in order to forget to think, to forget their concerns, their problems," as Stiegler similarly describes the chemical "hyperproletarianization" of youth created by drugs like Ritalin and Prozac.[66] Indeed, thousands of pharmacologists and biochemists were subsidized by the state in *Brave New World*: six years later, soma, the perfect commercially produced drug, emerged. Like Ritalin, soma renders a population hyperactive and attentive, calm and compliant. But it is not a drug for all classes. Epsilons have no need for soma; their brains have already been starved of oxygen at birth. This is a class system founded not only on good fortune, it would seem, but also on gradations of sensory deprivation and stimulation.

It is nonetheless perhaps Huxley's appeal to emotional conditioning that most significantly resonates with neurocapitalism. Although he makes a clear distinction between minds that desire and minds that decide, Huxley sees the advantages of a scientific education as a way to sidestep intellectual engagement altogether. Such an education system "ought never, in any circumstances... be rational."[67] The mind of the Brave New Worlder becomes the sum of all the emotional suggestions made to it, suggestions that guide intentions and subdue nonconformity. The College of Emotional Engineering is where Helmholtz Watson spends his time between lectures writing *feely scenarios* for radio shows, emotional slogans, and hypnopedic rhymes. Helmholtz's role as the emotional engineer is clearly inspired by the rise of mass media propaganda in the 1930s (Huxley locates the College of Emotional Engineering in the same building as the Bureaux of Propaganda),[68] but Huxley's unique focus on the emotional qualities of hypnotic suggestibility offers useful insights into the dystopian nature of more recent times.

FACEBOOK AND THE ENGINEERING OF EMOTIONAL DESIRES

For good reason perhaps, Nicholas Carr described it as a "bulletin from a dystopian future."[69] He was referring to an experiment Facebook carried out in 2014 involving the manipulation of the emotional content of news feeds and measuring the effect these manipulations had on the emotions of 689,003 of Facebook users in terms of how contagious they became.[70] The researchers who carried out the experiment found that when they reduced the positive expressions displayed by other users, they produced less positive and more negative posts. Likewise, when negative expressions were reduced, the opposite pattern occurred. Although the recorded levels of contagion were rather paltry, the researchers concluded that the "emotions expressed by others on Facebook influence our own emotions, constituting experimental evidence for massive-scale contagion via social networks."[71] Indeed, even if this contentious and ethically unsound attempt by Facebook to influence moods produced meager evidence of contagion, the design and implementation of the experiment itself should alert us to a potentially Huxleyesque mode of mass manipulation. As Carr further remarks, the bizarrely titled "Experimental Evidence of Massive-Scale Emotional Contagion through Social Networks" draws attention to the way in which the cultivation of big data assemblages by marketers treats human subjects like lab rats while also pointing to the widespread nature of manipulation by social media companies. "What was most worrisome about the study," Carr contends, "lay not in its design or its findings, but in its ordinariness."[72] This kind of research is indeed part of a "visible tip of an enormous and otherwise well-concealed iceberg" in the social media industry.[73] To be sure, the one thing that both the disparagers and apologists for Facebook seem to agree on is that user manipulation is rife on the Internet. It is, after all, what every social media business enterprise does. So why do we continue to share the intimate moments of our lives with these marketers?

Social media networks are clearly the perfect test bed, or nursery, for cultivating and triggering emotional contagions, because unlike the broadcast media, which also evidently spreads anxiety and joy, the users of these networks are predisposed, it would seem, to sharing their feelings in exchange for the tools that allow them to freely do so. They are, as Burroughs contends, a population who are happy to turn themselves on. Of course, despite the media storm surrounding this particular attempt to manipulate emotions, many users of Facebook will be joyfully oblivious of their participation in it and of the many other attempts to influence them, to grab their attention, or, indeed, of their own inclination to respond to emotional suggestion in such an apparently porous and imitative fashion.

Ostensibly, there is nothing particularly new to be found in these recent endeavors to produce attentive subjects and exploit the emotional desires of a population by making them contagious in order to steer intent. The history of marketing is littered with similar attempts to do so. St. Elmo Lewis's attention, interest, desire, and action model (AIDA), a prominent template for suggestive advertising developed in the late nineteenth century, made explicit the practical necessity to bring together desire and cognitive beliefs.[74] Similarly, in the 1920s, Freud's nephew Edward Bernays notoriously made the connection between attention, unconscious desires, and the selling of products to the masses in his marketing propaganda model. To be sure, the syllabus of any self-respecting emotional engineering degree must surely include a history of response and instrumental conditioning techniques, emphasizing the important role of Pavlov, Watson, and Skinner, but also bringing in Bernays's model to illustrate the efficacy of emotional manipulation in marketing practices. Indeed, Bernays well understood the leap from the mere conditioning of habitual responses and reaction psychology to a propaganda model founded on the creation of "circumstances which will swing emotional currents so as to make for purchaser demand."[75] As Helmholtz Watson might again recount in one of his lectures, the salesman who wanted us to "eat more bacon" would persuade us not because his bacon was the cheapest, or indeed the best, but because the doctor who recommends the bacon becomes a conditioning stimulus that feeds on a desire for authority. There would need to be, in other words, a complete circuit of Pavlovian control mechanisms in place to ensure at least some level of certainty that more bacon will be sold.

The difference today perhaps, as neuromarketers claim, is that the potential success or failure of these crude psychological marketing models can now be *guaranteed* by way of following the trajectory of the neurosciences. Marketers anticipate that these old models can be augmented by accurate biometric measurements of attention, emotional consumer engagement, and visceral stirrings that inform consumer choices. To be sure, since the mid-1990s, the neurosciences have gradually moved away from a purely cognitive-based approach toward an enquiry into the affective, emotive, and feely triggers assumed to be responsible for decision-making processes. The neuroscientific argument forwarded (the emotional brain thesis) suggests that the perturbations and disturbances of sensations elicited by certain feelings, like fear, can be subjected to response conditioning. There is an attempt, in the work of LeDoux, for example, to demonstrate how a rat's amygdala provokes a rapid response based not on cognitive but on emotional information processes.[76] Using Pavlovian conditioning, LeDoux points to a pathway that he contends fear travels through, from an input zone (the lateral amygdala) with connections to most other regions in the amygdala to the central nucleus, which functions as an output zone connected to networks that control fear behavior (freezing) and associated changes in body physiology (changes in heart rate, blood pressure, etc.).

Underpinned by neuroeconomic principles concerning the role of brain chemistry in financial decisions, neuromarketing attempts to go beyond a system of deciding that regards preferences as *a given* to explore the hedonic motivations exhibited by neurotransmitters thought to guide choice.[77] It is supposed that dopamine consequently updates the *value* an organism assigns to stimuli and actions, determining, some argue, the probability of a choice being made.[78] Like this, neuroeconomic propositions point toward the potential involvement of dopamine in the formation of expectations, beliefs, and preferences (assuming, that is to say, that expectations, beliefs, and preferences do not conversely affect dopamine activity).

If these kinds of neuroscientific suppositions concerning the processing of emotion have any credence at all, they will evidently challenge two canonical postulations at the heart of classical economics and persuasion theory. On one hand, the assumption that economic decisions are somehow guided by purely rational, utilitarian actors, rendered free from irrational emotions, becomes exposed to the uncertainties of a *reasoning* enmeshed somewhere in the networks that bring cognition and affect together. On the other hand, the emotional brain thesis also challenges the Machiavellian notion that fear is the most powerful means of social influence. As already discussed in the previous chapter, the question of what makes them click might be answered via appeals to a wide range of emotions. Similarly, a neuroeconomist might want to know what makes someone happy before a choice is made, because this state of mind can also greatly influence options.

Whether these neuroscientific ideas have any validity is beyond the remit of this chapter, but it must be noted that they have had a highly visible impact on business thinking. This is indeed a reciprocal affair. The work of prominent neuroscientific researchers is often cited in marketing literature, and many readily engage with the goals of neuromarketing.[79] There is an irresistible temptation, it seems, to draw on the neurosciences to grasp how attentive and emotional processing guides the purchase intent of a consumer in a supermarket or on a website. Neuromarketing aims, as such, to develop a fully operational model of persuasion that appropriates attention and emotional desires by way of conditioning reward systems and affective appeals. The objective is to make the stuff that motivates people to work harder, and consume more, predictable and therefore more efficiently reproducible in ersatz experiences. These affective, emotional, and feely appeals are part of a broader mechanism of persuasion encountered in neurocapitalism that attempts to manipulate the sensory environment by way of producing a stream of stimuli that conforms the mechanical habits of the worker and consumer to predictable, temporal behavioral patterns. Furthermore, this is a mode of persuasion quantized by the frequency-following forces of the market and assembled in the rhythmic entrainment of the brain–somatic relation, which, at the same time, paradoxically, transforms populations into attentive yet docile subjects.

The neuromarketers' readiness to plunder research into ADD and ADHD brings into focus the paradoxical relation capitalism establishes between attention and inattention. Indeed, it is this absurd continuum that tests the limits of control. Here is the crux of what Crary sees as a challenge to the normalizing powers of capitalism. Similarly, the capture of alpha brain time by popular industrialism in Stiegler's account means that attention deficit is, at once, a symptom of a disease and an insurgent cure. On one hand, the worst disease one can catch in the attention economy is not attention deficit. The problem is, of course, the overloading of attention. There is indeed never a lack of things to attend to. Accordingly, the *sickness* of inattention is not with the individual in deficit. It is rather a social relation with excess that is sick! On the other hand, then, inattention, distraction, reverie, and apathy all take on a potential therapeutic quality perhaps resilient to a control society that demands hyper attention and emotional engagement.

WHAT CAN A BRAIN DO?

Neurocapitalism needs to be understood as a combination of discursive and prediscursive forces. On one hand, and similar to the neuromarketing enterprise, the pharmaceutical giant has been complicit in linking the problem of attention deficit to wider economic discourses so as to intensify visibility, justify diagnosis, and, of course, increase sales. The intention is to highlight the "considerable costs associated with ADHD."[80] This extends Crary's pathologization of the inattentive child beyond the schoolroom into the wider economies that directly affect patients and their families, health care, social services, educational systems, the workplace, the criminal justice system, and even road traffic accidents, all of which are threatened by the costs of not taking the problem of ADHD seriously. Attention deficit becomes a discursively formed social disease similar in many ways to other syndromes located in the brain. This is a biopolitical positioning of ADHD that, similar to Schillmeier's recent account of dementia, becomes a national crisis, an irrepressible epidemic, a cancer that needs to be cracked.[81] Moreover, the neuropharmacy repeats the rhetoric of the neuromarketer; that is to say, following the economic fallout of the financial contagions of 2008, governments are struggling with economic deficits, so mental health problems like ADHD "are becoming less and less of a priority on the political agenda."[82] However, the discursive formations of big pharma tend to ignore concerns that austerity leads to more prescriptions for ADHD drugs because alternative care systems for unruly and inattentive children are more expensive.[83]

These discursive elements do not, however, circulate power relations in isolation. Psychostimulants like Adderall and Ritalin are increasingly implicated in the affective operations of the control societies. They have, as such, become more distributed, autonomous, and integrated into the systems of power. Indeed, to fully grasp these prediscursive forces encountered in times of neurocapitalism, it might be useful to once again engage with Fuller and Goffey's suggestion of a pragmatic approach to power.[84] Such an approach must begin by recognizing the collapsing of the biological and the technological into cultural and social ecologies. We cannot, for example, understand capitalism without taking into account its violent intervention into the rhythm of life. The mapping of brain wave frequencies to consumer experiences and neuropharmaceutical intercessions into behavioral, cognitive, and affective states are a deepening of this rhythmic intrusion. They are aspects of control that function according to oscillating rhythms intended to trigger frequency patterns of attention, inattention, arousal, and docility in the schoolroom as well as enforcing synchronizations and entrainments in the workplace and the shopping mall. From time to time, we are rewarded for being good pupils, workers, and consumers with the joyful dopamine rush. This is one way in which the rhythm of capitalism continues to motivate us to pay more attention and keep self-medicating. So how are we supposed to resist our dystopian future? Deleuze tells us, "There is no need to fear or hope, but only to look for new weapons."[85] Our choice of weapon depends to a great extent on what kind of resistance is capable of posing a serious threat to the joys of marketing and, at the same time, of cutting through the forces that motivate us to participate in these joys. It also means disturbing the rhythm of frequency-following capitalism so that we are able to fall out of its rendering of disciplinary time and take back control of brain frequency. It is my further contention that we need to grasp the terrible news that the corporation has a brain (as well as a soul) and turn it to our advantage, that is to say, to know not only *what can be done to a brain* but also *what a brain can do.*

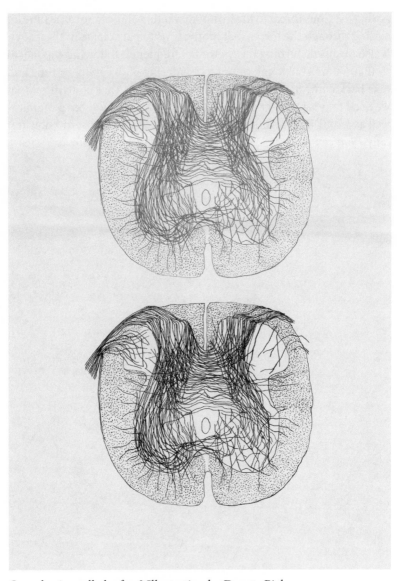

Can a brain really be free? Illustration by Dorota Piekorz.

PART II

What Can a Brain Do?

Before proceeding with the second part of this book, we need to note that the presence of Huxley is intended to provoke disciplinary interferences at all three levels (extrinsic, intrinsic, and nonlocalized). In the first instance, in *Brave New World*, we find a literary artwork that clearly resonated with the giants of science and philosophy. To be sure, the full force of *Brave New World*, according to the biochemist and sinologist Joseph Needham, could only really be "understood by biologists and philosophers."[1] Indeed, along with Needham, Bertrand Russell was a big fan.[2] Second, in terms of intrinsic interferences, Huxley's aesthetic figures enable some degree of slippage between planes of expression. Helmholtz Watson, Professor of Feelies, for example, allows for an interference between "weird" brain science and the marketer akin to the current potential of dystopic neuromarketing. As Needham points out, scientific knowledge is not immune to political interferences and must not be assumed to lead to a bed of roses:

> Huxley's orchid garden is itself an exemplification of the contention that [scientific] knowledge is always good, for had it not been for his imaginative power, we should not have seen so clearly what lies at the far end of certain inviting paths.[3]

Interestingly, Huxley's dystopia did not sit well with other writers of technofuturology. H. G. Wells describes *Brave New World* as a "betrayal of the future."[4] Furthermore, as the author of a cautionary tale of scientific progress, Huxley later adopted an ideological position that alienated many of his contemporaries and readers. As a result, he faced further criticism for being an escapist "who dodge[d] the troublesome job of moral and political decision by going up a rabbit hole and saying 'all is dark.'"[5]

Moreover, though, Huxley's eventful drift into a mescaline-induced philosophy of the Not-Self provides some impetus here for a non-localized interference between art, science, and philosophy and the concept of the assemblage brain. At this point, his writing certainly begins to resonate with Bergson's antilocationist stance in *Matter and Memory*. His continued interest in neurology, too (this time through an understanding of the effects of mescaline and LSD on brain chemistry), helps him toward a nonphilosophical, nonphenomenological experience of the *mind at large* that Deleuze and Guattari similarly sketch out in *What Is Philosophy?* It is indeed during one of his acid trips that Huxley eventually arrives at something akin to a Deleuzian microbrain state. His conclusion that it is not the "I" but the Non-Self that is involved in the process of perceiving the world (via becoming a chair leg) asks the same question as Bergson (and subsequently Deleuze and Guattari) concerning the limits of representational perception and the need to route around it.

However, there needs to be a couple of words of caution offered with regard to the limitations of Huxley's nonphilosophy of the Not-Self. To begin with, *The Doors of Perception* may well have been the book that launched a thousand trips, but any hope that it would prompt an emergent utopia was of course mostly lost in the dystopic haze of the late 1960s and early 1970s. The generation that Huxley apparently stirred into action were a *chaos people to come* who never really arrived. If they truly experienced a moment of utopia, then it was, perhaps, like *Brave New World,* little more than a dystopia in disguise. Second, his optimism about freeing the *mind at large* from the tyranny of representational perception maybe requires something more than an overly romanticized nonphilosophy of the affirmative. We must remember that the lines of flight of the rhizome brain can, and do, turn to deadly refrains. When Huxley died of cancer in 1963, just hours after Kennedy was shot, his partner injected him with a final fix of LSD. This deathbed trip was perhaps intended for him to decisively realize the Not-Self he so desired but most probably did little more than smooth out the pain of his death in a dystopic world.

Again, the imminent shadow of neuroculture will most probably be a dystopian affair. As part I of this book set out, the colonization of sense making by a digitally enhanced neurocapitalism threatens to be all pervasive. Nonetheless, the critical theory developed herein must not be afraid to dig itself out of Huxley's rabbit hole in the search for some light in the shadows. The assemblage brain is, as such, unashamedly, offered as a counter to dystopian media theory.

WAKING UP THE NEUROSOMNAMBULIST

To address the question of *what a brain can do,* the second part of this book follows, to some extent, Catherine Malabou's endeavor to free the brain, that is to say, in this context, liberate the brain–somatic relation from its coincidence with the passive ersatz experiences of neuro-capitalism. The ensuing chapters, as such, theorize an assemblage brain, an alternative emergent protosubjectivity distinct from the numerous attempts by marketers and pharmaceutical businesses, for example, to target a brain. These efforts are clearly linked to a trajectory of efficiency analysis beginning with Gramsci's observations of the eradication of the freedom to think nonconformist thoughts in the industrial factory and extends to the control of cognitive processes and, now, in times of neurocapitalism, noncognitive registers of affect. An assemblage theory of the brain must therefore realize a preexisting capacity for freedom and nonconformity in sense making. Like this, the discussion follows Malabou's effort to draw on neuroscientific theories of subjectivity, which suggest that although we might not know it, the brain is *already* free:[6]

> We are living at the hour of neuronal liberation, and we do not know it. An agency within us gives sense to the code, and we do not know it. The difference between the brain and psychism is shrinking considerably, and we do not know it.[7]

Significantly, the question of what a brain can do in the workplace, for example, is not answered, Malabou argues, by yielding to a model of plasticity redefined as flexible, that is, the flexible ergo docile worker, but rather requires a liberated plasticity that can know and modify *itself.*

Where this work significantly deviates from Malabou's thesis is in her contention that our brain is *us,* that is, that *We* coincide with "our brain."[8] To be sure, the search for a liberated protosubjectivity does not begin,

I will contend, with an "organic personality."[9] The assemblage brain is not a theory of individual freedom as such. The aim is not simply a matter of uncovering particular hidden brain regions and plastic processes in one brain so as to establish who we are or what we might become but rather, primarily, about awakening a collective political consciousness from its coincidence with the spirit of capitalism. Arguably, this collective freedom will not be achieved by looking *inside* the brain but rather by opening up sense making to the outside forces of relationality. Indeed, if we are to progress to "free this freedom,"[10] then it is better to ask how the brain relates to others rather than how it knows itself.

As we have thus far observed, the shadow cast by neuroculture is mostly a dystopic affair. Strangely, though, despite this, we find a dystopian trajectory that a large percentage of the population are, for reasons Huxley and Cajal's Mirahonda allude to, seemingly happy to go along with. Indeed, "happy" is possibly the operative word here, because this susceptibility to suggestion is not generally arrived at through cognitive processes alone but coincides to a great extent with affective, noncognitive encounters with the world implicated in processes of conformity and entrainment. As is the case in *Brave New World*, everybody's happy nowadays—*in everybody else's way*. Of course, other emotions are activated. But fear, anxiety, hatred, jealously, for example, are perhaps merely in the service of the subjectifications of Huxley's perfectly "happy, hard-working, goods-consuming citizen."[11] It is the pleasurable vices that lead to the degradation of the Brave New Worlders. As follows, Wilhelm Reich's quandary concerning why it is that the masses desire their own repression needs to be reconsidered,[12] or, to put it another way, and reworking Tarde a little, we need to ask why we encounter so much somnambulism in the world,[13] that is to say, a population seemingly sleepwalking into a dystopian political future by way of a hypnotic force of suggestibility. So, the liberating question of *what a brain can do* needs to be couched in such a way as to consider the sensory environments in which subjects become implicated in their own dire subjectification or *somnambulification*. What kind of subjectivity can be free enough, on one hand, to *think collectively* outside of the rhythmic time of the capitalist dystopia and, on the other, become sensitive to the sensory stimuli that produce somnambulist subjectivities, putting brains to work in so many unscrupulous and seemingly nonconscious ways?

It is important to note at this stage that I am not interested in discovering subjectivity or the emergence of a *self* in isolation. The Enlightenment concept of an emergent *selfhood* (Descartes's essence of human subjectivity) is firmly rejected. If the somnambulist is ever going to be woken up from this slumber, then I contend that there is a need to get to grips with the various processes by which protosubjectivity is supposed to emerge into a fully formed *selfhood*. However, significantly, rather than staying focused on the inner world of fully formed subjects, we need to grasp subjectification in the multiple processes of becoming. Thus far, I have opted to explore the brain-becoming-subject in the interferences that materialize when the brains of scientists, philosophers, and artists encounter chaos. Accordingly, the brain-becoming-subject is found in these encounters with infinity, and the subsequent planes cast by science, philosophy, and art might help us, it is argued, to comprehend subjectivity not as a ready-made person but as a process *in the making*.

In this second part, I begin by arguing that the assemblage brain will not be found in cognitive brain models, particularly those that lean on the engineering metaphors of representational storage and information processing as fundamental to the production of consciousness. It is important to note that the coupling of the cognitive subject to the cybernetic information machine is, above all, a political project rooted in capitalism. The cybernetic model does not free the brain but conforms it to the rhythmic frequencies and entrainments of digital capitalism. The cognitive subject is indeed the antithesis of the nomadic freedoms of the assemblage brain. The cognitive subject is a fully formed brain-subject connected to its environment via encoded and decoded flows of information. Cognitive subjectivity is a territorialization of consciousness limited to a mind that imitates conformist opinions and received ways of perceiving the world through the prism of capitalism. In contrast, the assemblage brain needs to be free to route around the imitation of public opinion and common sense. Consequently, the preconceptions of the somnambulist need to become more deterritorialized and politically attuned to the sense-making relations established between brains and sensory environments. The theory of the assemblage brain must be able to discern between environmental stimuli, which might, on one hand, control by way of producing docile subjects and, on the other, empower by activating and freeing up collective brain time.

Moreover, to route around the fixed opinions and perceptions of fully formed brain-subjects, there needs to be an understanding of a brain-becoming-subject established through imitative relations, panpsychic and affective encounters with an environment inclusive of nonhuman and inorganic matter.

To begin to theorize the assemblage brain, this discussion will continue to explore the potential of interferences between philosophy, art, the neurosciences, and capitalism. For example, through its influence on such things as neuromarketing and neuroeconomics, the emotional turn in the brain sciences has provided capitalism with new insights into how to exploit the sensory environments of a population, but this knowhow about the interwovenness of affective and cognitive ecologies might also, inversely, contribute to an understanding of brain emancipation. In terms of philosophizing the sense-making processes of Tardean somnambulism, there is much to learn from an approach that evidently provides valuable insights into the noncognitive emergence of consciousness. Beyond that, art can, like this, provide further valuable interferences that tap into the somnambulist's immersion into the noncognitive ecologies of neurocapitalism. The artist is, after all, well practiced in the production of the kinds of sensations that expose and critique the sensory environments in which neuropower subsists. Indeed, protosubjectivity is perhaps better understood by way of the artist's sensations than it is through the interrogations of the philosopher's concepts. This is not to say that the artist has a special, or distinctly different, *brain type*. On the contrary, this is an antilocationist stance in which it is the artistic plane of expression, with its added affect, that can make art potentially dangerous.

TARDE'S SOCIAL BRAIN

Ultimately, the assemblage brain will need to escape from the *dream of action* that constitutes the tyranny of Tardean somnambulism. This is the reverie in which the sleepwalker comes under the influence of an *action-at-a-distance,* an attraction to points of fascination and intoxicating celebrity glories that intensify vulnerabilities to persuasive suggestibility in the sensory environments of neurocapitalism. So where are these productive interferences to be found in Tardean somnambulism?

Following Tarde, we can see how his mind contagions are composed somewhere in between desire and belief, just as they are today, similarly, established between affect and cognition in the neurosciences. Unlike Durkheim, who tries to grasp the social as distinct from psychology and biology, Tarde provides an "understanding of social 'associations' with no distinction made whatsoever between Nature and Society."[14] "Nothing is less scientific," Tarde argued, than the establishment of this absolute separation, of this abrupt break between the voluntary and the involuntary, between the conscious and the unconscious.[15] Do we not pass by insensible degrees, he asked, from deliberate volition to almost mechanical habit? The somnambulist subject "unconsciously and involuntarily reflects the opinion of others, or allows an action of others to be suggested to him."[16] The sleepwalker is caught in an intersection between a culture of attraction and a biologically hardwired inclination to imitate.

In many ways, then, Tarde's appeal to the in-between of desire and belief offers a neurological perspective on subjectivity in the making: a production of a porous relation with the sensations and concepts of others established between the cells in different brains. Indeed, the concurrence of Tarde's imitation thesis with a contemporary understanding of "a brain circuitry that fires when we either perform a given action or see someone else perform the same action" has already been noted.[17] Not unlike Deleuze, then, in his time, Tarde demonstrated a deep commitment to early neuroscience by drawing on the British psychiatrist Henry Maudsley's theory of an innate tendency to imitate in the nervous system as a basis for understanding how one brain could fascinate and polarize the desires and beliefs of another. Nevertheless, Tarde's imitative brain is not grasped here as genetically determined or neurocentric. A Tardean assemblage brain is in fact a *social brain* that collapses the distinction between biology and culture. It is a brain in which biological cells are etched with social traces of the other. So Tarde's theory of desire and belief refers to cerebral imitation functions, but these are not significantly located *inside* the brain. They are, rather, understood as a brain *reaching out* to the social world.

As such, a number of considerations need to be taken into account before going on to discuss how Tarde's appeal to panpsychism can supplement assemblage theory as well as grasp his contribution to a more contemporary understanding of radical imitative relationality:

1. The imitative associations that relate one brain cell to another are not a deeply hidden secret. They are not unconscious or disconnected from the world. Tarde's belief and desire are not, as such, located in the deep recesses of the brain but rather in social encounters wherein nonconscious states overlap conscious states. This is where the dreams and woken states of the somnambulist become indistinguishable.

2. Nonconscious associations are not structured like a language either. That is, before language, the social brain reaches out to *outside things* by way of imitation. There are indeed interesting intersections between this interpretation of the Tardean brain and Bruce E. Wexler's more recent work on the interactions between the brain and culture, which similarly contends that language is not a property of the brain but an effect of imitative sensory environments. As Wexler contends, if all living people were rendered permanently speechless and illiterate, their offspring and succeeding generations would be unable to speak despite having normal brains, and language, this most distinctive of all human characteristics, would be lost to the human species.[18] It is a deep ontological commitment to imitation, not language, which becomes the universal principle of the social brain. As Wexler puts it, imitation is "consistently operative throughout the moment-to-moment unfolding of everyday life."[19]

3. The defining characteristics of the human brain need to be more closely aligned to the imitative capacities of other animals, who seem to understand one another, Tarde argued, almost without signs, as if through a kind of infracommunication of suggestion. For example, birds do not appear to think in symbols or words but nevertheless form elaborate modes of communication founded on their seemingly shared capacity to imitate. What spreads in bird communication *reaches out* to others (and other things), borrowing desires and producing multiple territorial arrangements of belief. Like Proust's fat bumblebee fertilizing the orchid, these microarrangements are always made *outside* of the species line.

4. It is among these microarrangements of birdsong that the *ritornellos* and *lines of flight* of assemblage theory clearly intersect with the Tardean brain. Both Deleuze and Guattari, and Tarde,

considered the refrain of birdsong as an illustration of the important distinction that needs to be made between the social and the genetic brain.[20] According to memetics, all efforts made by animals to be social in the human world are abortive due to a genetic failure to evolve imitation into advanced cognitive processes like language. But every animal, like every human, reaches out to the social life according to its innate capacity to imitate. This is the sine qua non of Tarde's neurological development: a precondition of all social life. So unlike the memetic bird, who imitates the songs of its species line so as to secure genetic territorial boundaries, Tardean birdsong is often complicated by what appears to be the many examples of cross-species imitation. It reaches out and borrows from interconnecting lines of communication. There is indeed a "deep-seated desire to imitate for the sake of imitation."[21] Birds imitate other birds so precisely that they are often deceived. They can imitate humans or tape recordings of other birds. They also mix their own notes with those of other species nearby. Birdsong becomes an escalating occupation of frequencies. Guattari, who was incidentally a pianist, grasps the musicology of this imitative occupation—how the return to a repeated theme brings together the singularities of an improvisation, and how the repetition of imitation brings unity to composition. But the more birds, the more the species lines, get crossed, and the more lines of communication get crossed, the more the refrains are exposed to the outside. The society of imitation becomes a multiplicity defined by the outside: by the abstract line, by the line of flight.

5. Tarde makes a very important distinction between the felt sensations we might experience in a cinema, for example, which simultaneously affect the attentive crowd, sending a shudder up their collective spines, and the *in-between* of desires and beliefs. Crucially, Tarde was not opposing a theory of sensation to a theory of concept. As Bergson similarly contends, there are insensible degrees between, on one hand, affective states and, on the other, representational states.[22] It is this *in-between* of relationality, or the interference, as I call it here, that must remain the focus, because it is in the intermediacy of desire and belief where intensities may become socially contagious.

6. Finally, the question of what a brain can do is, in part, inspired by Tarde's concept of three kinds of individuality.[23] Significantly, the organic, the psychological (the "myself"), and the social individual are not grasped as self-contained entities at various levels of a layered social model but are all considered in relation to their duration in imitative assemblages. These are imitative relations established and maintained between cell-to-cell transmissions of resemblance and variation, shared external sensations, feelings, beliefs, desires, suggestibility, and influence.

ROUTING AROUND PERCEPTION AND OPINION

Developing this account, the second part of the book conceives of a notion of subjectivity in the making (protosubjectivity) that is not located *inside* the brain. On the contrary, the assemblage brain establishes a radical relationality between the individual and the sensory environment wherein sense making does not proceed on a journey from the inside to the outside, or vice versa. The inside is nothing more than a fold of the outside.[24] It is also not a fully formed subjectivity, that is, the self *we* perceive of through a language that tells *us* who *we* are. The imitations of the assemblage brain are prior to language and its grip on perception. To be sure, without imitation, language would surely never have thrived, and we would have no words to describe who we are. That kind of sense making would perhaps just be a sensation.

This notion of sense making is also deeply political and pertinent to an understanding of neurocapitalism, which has started to exploit the noncognitive, affective realms of protosubjectivity. The assemblage brain thus needs to be antagonistic to Enlightenment notions of self and the Enlightenment's legacy in neoliberal concepts of free-willed subjectivity, which are continually dividing up individuality into smaller exploitable units like those located in the mental stuff of the living neuron.

Tarde found some of the radical relationality of the assemblage brain still evident in the animal-like behaviors of the crowd, which seems to become magnetized by the uncanny sensations of imitative radiation. However, he also notes how the newly mediated spaces of the industrial age introduced far more docile publics. Tarde's notion that the nonconscious associations of the crowd could be steered in such a way as to make them even more stupefied and open to imitation and suggestibility is surely a precursor to a media theory viewpoint that would eventually go on to argue that there is no such thing as the public, let alone public opinion.[25] This is Tarde's *dream of action*—a political nonconscious made up mostly of mechanical habits and hypnotic relations with others, composed in such a way as to appear that the concept of "myself" has emerged as a fully formed and coherent subject with a sense of its own volition as an individual and as part of the public. This is Tarde's *illusion of personal identity*: the somnambulist. This is not a "myself" that is completely unconscious but is like someone who, on waking in the morning, "passes by degrees from the dreaming 'myself' to the 'myself' awakened."[26]

The question is, how can this emerged sense of self we perceive of, and the wider sense of public opinions that seem to belong to it, be routed around? Indeed, to wake up the somnambulist is to get to *know* the assemblage brain, and to do that, we need to grasp that there is no self that is not etched with the sensations of the other or the sensory environments in which we encounter the other. To make sense of the assemblage brain therefore means that we have to think the unthinkable or, to put it another way, think in sensation.

Selected frames from Milos Rajkovic's (aka Sholim) 命.

4

Sense Making and Assemblages

SHOLIM'S THUMB

In the first part of this book, I noted a prevailing trend for human brains
to be grasped metaphorically, by scientists, philosophers, and artists,
as computers. This is a general cognitive paradigm in which individual
brains have been conceived of as hardware systems from which a cognitive mind emerges like a software package. This is the computer–mind
metaphor, or the cybernetic brain model, still favored by many cognitive
neuroscientists. When considered collectively, brains similarly become
part of a distributed human–machine information network, emerging as
collective forms of intelligence or consciousness, as initially advocated
by cyberculture theorists back in the 1990s. Following an interaction
between the cognitive parts of a human brain, like memory, and the
parts of a technological network, a threshold appears to be met at which
human intelligence becomes the sum total of who *we* are in the digital
age. This is a cognitive prostheses: the much-celebrated *megamind*.[1]
   Serbian artist Milos Rajkovic's (aka Sholim) 命 offers a refreshingly
alternative perspective on the megaminds of digital culture. Through
a series of Magritteist looping GIF animations, he manages to capture
the rhythmic aesthetic of a less celebratory age of the scrolling thumb.
Forget the utopias of cognitive prostheses or collective brain ecologies; as Gary Genosko cannily points out, "the digital world is well on
the way to becoming all about thumbs."[2] Accordingly, Sholim's thumb
not only draws attention to the role of the hurried thumb scrolls of
digital cultures but also offers insight into the noncognitive relations
established between human body parts (other than brains) and their
precious smart phones. Gone are the cognitive ecologies of brains

and computers. The emergence of collective intelligence is replaced in Sholim's art by a population connected up to each other via the hollowed-out machine heads of popularist politicians, corporate workaholics, and superficial celebrities. Indeed, 命 (Chinese for "life")[3] well captures the emergence of noncognitive ecologies of body parts that sense the digital world through haptic rather than purely cognitive associations. As a consequence, 命 opens up significant political questions concerning the emergence of a noncognitive megamind, because the action of hastily scrolling past the content becomes trapped in a short video loop of an exploited Chinese worker bolting together the components of the smart phone in some dismal factory. Within the recurring scroll of the thumb, Sholim allows the viewer only a fleeting moment to reflect on these tyrannies of the digital age. The sensation he offers inspires a number of questions. For example, what kind of uncaring megamind is this in which the worker toils as thumbs continue to scroll? What kind of collective consciousness (or nonconsciousness) is it that skims over the harsh actualities of worker exploitation to get to the entertainment? What kind of sense-making assemblages make the thumb complicit in these abuses? What kind of haptic sensation is the thumb experiencing that makes it so seemingly numb to the world? Is this the emergence of a new kind of rhythmic collective nonconsciousness in the digital age—not a megamind at all but an assemblage of body parts in which thumbs scroll, workers grind, and capital churns?

The overall aim of this chapter is to use assemblage theory to re-imagine the emergence of *subjectivity in the making* as a constituent of an all-encompassing, panpsychic relational process of sense making, inclusive of workers and thumbs, humans and nonhumans, organic and nonorganic parts. This is an understanding of an inclusive proto-subjectivity that also needs to be sensitive to, and considerate of, the wider sensory environments these component parts populate, including human labor and natural resources, like tin, that are exploited and plundered so that thumbs can continue to scroll. To contemplate the politics of these assemblages of sense making, a fundamentally different concept of emergence is required, particularly different from those apparent in cybernetic models of the brain that have informed cyber-cultural notions of collective consciousness since the 1990s, wherein collectively distributed brainpower is conceived of as somehow capable of supervening individuated component parts. That is to say, despite an ostensible commitment to a hardwired material brain, what emerges in these accounts of the megamind becomes a transcendent, unified whole—a sum total of who *we* are collectively—found mysteriously floating above the inner, microlevel interactions of brain matter. Here, in contrast to the cybernetic model, I work toward an assemblage theory of sense making that does not distinguish between a miraculously formed macro-outside and the microinteractions found inside but instead points to the role of intermediate relationalities in processes of emergence.

My approach also encourages a very different kind of emergent *neurodiversity* to that forwarded by models that insist on retaining the rational human brain at the center of an Enlightenment universe. Where, after all, is the sensation of the thumb in this brain-centered universe? Indeed, we will need to begin by briefly revisiting the claims of certain theorists who, in the late 1990s, borrowed from cybernetics to rejoice in the coming together of brains and digital network technologies as an emergent rational collective consciousness. Although often celebrating its posthuman credentials, much of this theorizing amounted to little more than an exclusive extension of humanist ideals, such as reason, progress, and brain faculty, limited to an anthropological notion of selfhood. Cyberculture was indeed considered a "legitimate heir" of the Enlightenment project.[4] Perhaps we should not be so surprised that what has emerged from these interactions is, for the most part, an expansion of the capitalist political model critiqued by artists like Sholim.

My theoretical objective in this chapter is twofold. First, I want to open up these human-centered models of collective consciousness to the interferences between science, philosophy, and art, and second, I look beyond the problems encountered in the human-centered modeling of collective consciousness toward a more inclusive notion of how matter becomes able to make sense of the world. Accordingly, the chapter is structured around the critical evaluation of three models of the brain. The first two follow fairly recent and contrasting theoretical trajectories in the neurosciences intended to shake the foundations of the cybernetic model. I begin with the *synaptic self*, which challenges the Enlightenment notion of selfhood by considering emotions, affect, and feelings as enmeshed in the material processes of reasoning. Indeed, this model becomes central to Catherine Malabou's effort to liberate the brain from cybernetic rigidity, genetic determinism, and capitalism by way of realizing the potential of neuronal plasticity. Nonetheless, the freedom afforded to the plasticity of the synaptic self has been further confronted by another brain: *the systematic self.* What constitutes selfhood in this model, on one hand, notes the failure of the synaptic self to break from both Cartesian dualism and the information metaphors of the cybernetic brain and, on the other, upholds the inclusivity of body parts, other than brains, in the emergence of subjectivity. There is, it seems, a place in this system of sense making for thumbs—even big toes!

The chapter concludes by bringing in a third brain, conceived of through a combination of assemblage theory and Tarde's monadological panpsychism and intended to draw attention to a number of flawed assumptions that buttress the two previous models. In a word, although opening up the political potential for plasticity and new interacting body parts, both the synaptic and systematic models produce a unified whole that miraculously supervenes all individual parts. This is a problematic mereological tendency that has persisted in emergence theories dating back to the ancient constitution of the problem of how the *One* emerges from the *Many*—how, that is, a multitude of parts comes together to form transcendent, stable wholes. This emergence philosophy follows a trajectory stemming from the ancient Greeks, passing through the Enlightenment to the birth of social theory in the nineteenth century, and enduring into present-day conceptions of the megamind. A Tardean intervention is indeed crucial here because the kind of cybersocial emergence theory apparent in the late 1990s has a distinct (and problematic) Durkheimian slant to it.[5] That is to say, it presents a collective consciousness that emerges as the highest level of human physic life. The quiddity of what Durkheim called the *consciousness of consciousness* is squarely located *outside* and *above* individual and local contingencies. Similarly, and despite locationist tendencies, in the synaptic and systematic manifestations of the self theorized in the neurosciences, we find a global brain that floats conspicuously *over matter,* producing enough downward pressure eventually to constrain and determine the interaction between parts. Both individual and collective cognitive models similarly declare that the whole is erroneously greater than the sum of its parts. In contrast, an assemblage theory of protosubjectivity does away with wholes altogether, favoring instead a focus on the sometimes puzzling interactions between parts as key to what brings components of sense making into relation with each other.

CYBERNETIC CONSCIOUSNESS

As a response to the dominance of what Arthur Kroker has called the "panic Enlightenment" model of the brain, which "limits consciousness to individual consciousness, reducing the brain to an organ trapped in a physical cranium,"[6] the emergence of collective consciousness seemed to offer the disaffected posthuman philosopher a potentially magnificent coupling of human brains and computer networks. Since the mid-1990s, these theorists speculated about the emergence of networked *thinking communities* or *cognitive ecologies* that were so crucial to what it meant to *become virtual* in the digital age.[7]

Prominent among these cybernetic emergence theorists at the time was Pierre Lévy, who conceived of a "self-organized emergence" based on a combination of "biological possibilities, cultural forms, social networks and intellectual technologies."[8] Lévy's approach stands out because, unlike many other cybernetic theories of emergence, his megamind is structured around an inclusive, associative topological universe of relation; a cognitive ecology or "a different order of reality" interwoven with feelings and emotions as well as thought and ideas.[9] However, what emerges from these biological, cultural, social, and technological interactions does not entirely escape the Enlightenment notion of a brain. As mentioned earlier, the megamind was conceived of as a successor to the Enlightenment project.[10] Along with other optimistic cybertheorists at the time, Lévy hoped for a unifying cyberculture—an emergent, transformative, and benevolent global brain trust. The onset of the digital network in the 1990s seemed destined to change forever the relation between technology and brains, that is to say, bringing together an interaction of human and technological parts that would trigger a novel electronic virtualization of collective intelligence.

It is much easier, with hindsight, to observe the many inconsistences in these early cybernetic models of emergent collective consciousness. To be sure, what has emerged in cyberspace today seems very distant from the much-anticipated digital revolution of the cognitive mind. Conceptually speaking, the mismatch between the neoliberal extension of Enlightenment thinking and posthuman dreams was always going to run into problems.

THUMB BRAINS

The application of cybernetically inspired models to explain emergent consciousness also missed two fundamental points about biological, cultural, social, and technological relations. First, and aside from Lévy's inclusion of an affective dimension, most models overlooked the significant role of affective brain states implicated in the coming together of humans and technology. By prioritizing conscious cognitive brainpower, cybercultural theory missed the opportunity to glimpse the emergence of *noncognitive ecologies*. The brain, as Kroker points out, has always been deeply affective. Indeed, before brains became digital, they were already well intertwined in their own affective bionetworks. There is, like this, a "complex intermingling" of gut brains, vision brains, tactile brains, ear brains: brain matter connected to the "full human sensorium."[11] This is indeed a neurodiversity that does not figure in the cybernetic emergence of collective consciousness: "the fact that our brains are not one but multiple, not trapped in craniums [or computers] but essentially relational" to each other in affective networks. To be sure, the desires necessary for an event like the Arab Spring to occur were never going to be produced by a purely rational cognitive ecology. Like all revolutionary social movements, the Arab uprisings were contingent on a collective sensorium of affective transmission; a collective nonconscious, in Tardean terms.

Second, theories of cybernetic collective intelligence have largely ignored an assortment of other interacting body parts implicated in the emergence of mutual sense making. In addition to the synaptic fusing of brains and digital networks, there is a need to therefore consider the vast assemblages of other body parts enmeshed in the emergence of this collective nonconscious. Indeed, whether it is the thumb or Genosko's much-favored big toe,[12] we seem to be encountering a relation between diverse body parts that have been previously neglected in theories of emergent collective consciousness. The tweeting, Facebooking crowds that made up the Arab Spring were clearly composed of a swirling mass of thumbs, but with the onset of wearable technologies, which turn shoes, for example, into computers, such interactions are being extended to big toes.[13] So, in many ways, the inclusivity of body parts, other than brains, begins to reorient the question concerning *what a can brain do* toward a more comprehensive model of diversity, inclusive of the sense-making capacity of the *thumb brain*.

The inclusion of the affective capacities of other body parts must not, however, be seen as a panacea to the troubling dark side of digital culture. As Sholim's thumb demonstrates, all kinds of body parts become affectively related to each other in the post-Taylor endeavor to root out the evil of worker and consumer inefficiency and nonconformity. The scrolling thumb brain is, perhaps, best understood as nothing more than an extension of Gramsci's muscular labor into the post-Taylor factory and the related onset of new technological paradigms, such as the Internet of things. The introduction of wearable technologies to the digital workplace and shopping mall brings all kinds of body parts (brains, fingers, toes, thumbs, wrists, ankles, etc.) into a controlling relation with computing, making them ripe for efficiency analysis and conformed to the increasing rhythmic speed of the capitalist machine. The imminent arrival of ubiquitous computing certainly forces us to rethink what kind of collective sense making emerges when body parts mix with, or take a backseat to, autonomous networks and artificial intelligences mediated through intangible interfaces. Indeed, the rapid expansion of the technological unconscious raises big questions concerning the politics of noncognitive ecologies. The assemblage brain, via an updated reading of Tarde's panpsychism, needs to encompass parts that are nonhuman, as such. Indeed, as ubiquitous computing spreads to such things as fridges, shoes, and lampposts, human individuation cannot be discounted from the pervasiveness of silicon circuits, Wi-Fi, and relational databases. But before these ecological relations can be further contemplated, the discussion draws on the neurosciences to explore how the emergence of a collective nonconscious might be conceived of.

THE SYNAPTIC SELF

Joseph LeDoux and Antonio Damasio have similarly contended that the coherent sense of self individual humans experience is an emergent outcome of nonconscious interactions located at the microlevel of synaptic functionality. This notion of a synaptic proto-self is of interest here because (1) it returns the discussion to Deleuze and Guattari's tendency to lean on the brain sciences, in particular a materialist brain in which nonconscious processes are contiguous with the emergence of subjectivity in the making, and (2) it suggests a diachronic emergence from the materiality of nonconscious neuronal states to conscious mental states, grasped mainly through the concept of brain plasticity. On one hand, these neuroscientific theories, which ostensibly break with the cybernetic model, produce productive interferences with a Deleuzian ontology in terms of developing a material theory of sense making. However, the plastic brain thesis also introduces significant problems with regard to an understanding of the emergence of self as a *sum total* manifested in the interactions between brain parts. On the other, the plasticity of the synaptic self informs Malabou's significant intervention into the neurosciences, in which she points to the potential for a political notion of sense making grasped in relation to the sensory environments of capitalism. Like Tarde, to some extent, this plastic brain becomes a junction point where biological, social, cultural, and political domains collapse into each other. However, I further contend in what follows that although astutely critical of the apolitical nature in which the synaptic self is presented in the neurosciences, Malabou's eventual commitment to it misses the problematic retention in this model of the metaphors of information processing and representational storage inherited from cybernetics. Indeed, through the intervention of Henri Bergson's nonrepresentational brain from *Matter and Memory,* I begin to unpack some of the problems that persist in these metaphorical renderings of sense making.

The Sum Total of Who We "Think" We Are?

Deleuze and Guattari's chaos brain appears to concur with the synaptic self insofar as both speculate that it is not the mind or the person but the brain that thinks.[14] To be sure, in both models, we find a brain that can think itself, a kind of self-awareness that might, one day, be explained by the materialist perspectives of the neurosciences.[15] Indeed, the synaptic brain would seem to coincide with Deleuzian ontology insofar as it also endeavors to avoid the trappings of the Cartesian theater, heading instead toward a *sense of self* emerging from the electrochemical activities of the protoplasmic mass of a brain, while also steering away from the tired old computer–mind metaphor of the cognitive turn.

Nonetheless, taking into account Bergson's significant influence on Deleuzian ontology, questions arise concerning the journey between the microlevel of neuronal interactions and the emergence of a stable sense of self. As Bergson deftly puts it,

> that there is a close connection between a state of consciousness and the brain we do not dispute. But there is also a close connection between a coat and the nail on which it hangs, for, if the nail is pulled out, the coat falls to the ground. Shall we say, then, that the shape of the nail gives us the shape of the coat, or in any way corresponds to it?[16]

Before further exploring Bergson's contention, it is necessary to set out the synaptic self hypothesis. One strength that needs to be noted is its inclusion of a noncognitive dimension to emergent and adaptable subjectivity. Nonetheless, this comes at a price, because it requires an ontological commitment to a notion of emergence resulting in parts coming together to form a sum total. So, at first glance, LeDoux's synaptic self, and Damasio's proto-self, are interesting proposals insofar as they do not cast out nonconscious processes into the wilderness of Descartes's animalistic physical world. They do not assume, as such, that a brain is immune to somatic affective experiences. This means that the concept of self is not a composition of purely explicit cognitive functions (perceptions, attention, memory, etc.) but emerges from mostly nonconscious experiences of the world. This not only challenges Descartes but also substitutes the Durkheimian emergence of the consciousness of consciousness with we might call here *the consciousness of the nonconscious*. Indeed, the emergence of the synaptic self is contingent on the historical processes of plasticity, which allows the brain to both learn and be modified by sensory associations made in an environment. In other words, the neuroscientific self may start from the same perspective of Descartes, that is, seeing the mind as the product of the brain, but in contrast to philosophies obsessed with substances and logical relations established between matter and mind, the synaptic self goes back to the material basis of neuronal functioning, trying to understand how the brain makes the mind possible.

Although, typically, we end up with a totality of mind over matter that conflicts with the open-ended nature of Deleuzian ontology, it is important to note that this sum total is not only ascribed to processes *we* become aware of. The *whole* of the synaptic self (who *we* are; the *I*, the *me*, or the *person*) is attributed to the nonconscious production of brain processes that "allow cooperative interactions to take place between various brain systems that are involved in particular states and experiences."[17] In short, LeDoux's sense of self (the *essence* of who we are) is an emergent whole that results from microlevel synaptic interactions and patterns of interconnectivity between neurons in the brain. What emerges is a "totality of the living organism."[18] Following a refined version of William James's notion of consciousness, LeDoux's self becomes a "stored" sum total of everything we are "in and between the various systems of your brain."[19]

Damasio's proto-self offers a similar series of complex interactions between brain parts, bringing about a sense of a coherent whole; that is, a brain that holds within it a model of the whole self. This is, however, grasped through an asymmetrical and representational relation between parts. Some parts of the brain, Damasio contends, "are free to roam over the world" and *map* "whatever object the organism's design permits them to map."[20] In contrast, other parts of the brain, which represent the state of the organism, get *stuck* and are unable to freely roam, mapping nothing but the body, and do so within largely preset maps. This makes them a "captive audience [and] at the mercy of the body's dynamic sameness."[21] The proto-self therefore emerges as an outcome of these irregular relations that, on one hand, remains relatively stable throughout the qualitative lifetime of an organism (its composition and general functions), while, on the other, small quantitative bodily changes constantly emerge within a limited range of parameters, allowing the organism to stay constant and consequently survive in the surrounding environment.

Like LeDoux's synaptic self, the proto-self is both stable and historically formed, due, Damasio argues, to an elaborate neuroanatomy made of many parts that nonconsciously govern the stability of this sense of wholeness. Like this, the neural machinery senses small variations in the parameters of the body's internal chemical profile and acts to adjust these differences, directly or circuitously. So consciousness emerges, according to Damasio, as an *entire unit,* that is to say, the living being, or body. But it is a part of this entirety called the brain, nonetheless, that "holds within it a sort of model of the whole thing."[22] The proto-self takes the form of an *organic representation of the organism itself* contained in the brain that maintains coherence. This representation of all other parts of the organism is what Damasio considers to be the most likely "biological forerunner" of the sense of a "preconscious biological precedent."[23] From out of this proto-self emerges a series of levels, beginning with the core self, or *core consciousness,* or *I,* and the temporal and historical permanence of the subject expressed in an autobiographical self containing the "invariant aspects of an individual's biography."[24] The interactions between these parts of the brain are, for that reason, deeply rooted in more elaborate representations experienced as identity and personhood. The consciousness of the nonconscious thus emerges like a coherent matryoshka-like network of maps without rupture or leap.

The Plasticity of Self

Significantly, the emergence of a neuronal sense of self is, according to LeDoux, a unity, but *it is not unitary.*[25] The plasticity of the synaptic self is not, as such, a personality formed out of a genetic imprint. It is "added to and subtracted from"; it is a plasticity of "genetic maturation, learning, forgetting, stress, aging, and disease."[26] This plasticity is what Malabou importantly draws attention to when she asks the question of what we should do with our brains. We should not be concerned about a genetically determined intelligence because the plastic brain provides a "possible margin of improvisation" between the synapse and the biological encoding of genetic necessity.[27] "Today it is no longer chance versus necessity, but chance, necessity, and plasticity—which is neither the one nor the other."[28] Nonetheless, despite the neuronal personality having a historical temporality, the improvisation between gene power and variability is grasped by Malabou as a dialectical "synthesis of all the plastic processes at work in the brain," allowing the organism to "hold together and unify the cartography of networks."[29] Indeed, according to the synaptic self hypothesis, the emergence of stable thoughts, emotions, and motivations becomes necessary for survival of a coherent and rational neuronal personality; otherwise, irrationality would cast these thoughts out into the wilderness, and emotions and motivations would be scattered in all directions like some unruly mob.[30] It is the emergence of this desire for stability that permits the coherent group-ing and linking of thoughts, emotions, and motivations in the synaptic self and proto-self.

While not entirely static, then, the proto-self and the synaptic self are, nevertheless, composed of a *totality* of everything the organism is physically, biologically, psychologically, socially, and culturally. So what initially appears to be an absolute rejection of Cartesian dualism and installation of a purely materialist brain needs to be more cautiously approached. This is because, in the midst of this seemingly materialist perspective, there is a tendency to unconvincingly exorcise the ghost of Descartes; Cartesian structures keep creeping back into the concept of an emergent consciousness. This perspective certainly presents a mereological puzzle that, by and large, sits uncomfortably with the materiality of Deleuze and Guattari's expression of neurophilosophy. The details of this mereological puzzle will fully unfold in the following pages, but ahead of trying to solve it, we need to briefly make note of a series of related problems concerning the materiality of the neuronal self, primarily, the representational and dialectical nature of the proto-self and the apparent failure of Damasio and LeDoux to completely escape the computer–mind metaphor, but also, importantly, addressing Malabou's observation of a political deficit in the neuroscientific model of the synaptic self.

The Problem with Representational Thinking

To begin with, what are we to make of Damasio's representational emergence of self? The emergence of the necessary stability of a fully *mental* sense of self is, according to the proto-self model, arrived at recursively through a series of representations commencing at the microlevel of neuronal patterning. Here we need to initially consider Malabou's acknowledgment that such a concept of self flies in the face of a Bergsonist nonrepresentational position concerning what constitutes the journey between the coat and the nail it hangs on, that is, between matter and mind. As Malabou explains,

> there *actually* is, contrary to Bergson's claim, a self-representation of the brain, an auto-representation of cerebral structures that co-incides with the auto-representation of the organism. This internal

power of representation inherent in neuronal activity constitutes the prototypical form of symbolic activity. Everything happens as if the very connectivity of the connections—their structure of reference, in other words, their semiotic nature in general—represents itself, "maps" itself, and precisely this representational activity permits a blurring of the borders between brain and psyche.[31]

This disparity between the *semiotic nature* of Damasio's self and Bergson's nonrepresentational brain deserves a much closer inspection. In *Matter and Memory*, Bergson's concept of the virtual, or movement image, famously exceeds what the idealists would call a representation, while at the same time not altogether siding with what realists would call a thing.[32] The virtual image is an existence placed halfway between the *thing* and the *representation*. When we encounter a novel object, Bergson contends, we are not experiencing it as a representation already *stored* in the brain, and likewise, if we have encountered the same object a thousand times before in the past, our experience of it does not become a representation stored in memory. Our memory of things becomes a mechanical habit passing from the original stimulus to the brain without the "miraculous power of changing itself into a representation of things."[33] Conscious processes, such as those experienced in recollecting the past, but also, arguably, these seemingly totalized experiences of self and identity, are not preserved in the brain. They are *preserved* in themselves—*in their duration*. Bergson's matter therefore becomes an aggregate of *images,* or the intermediary existence of the thing and the representation.[34] Importantly, then, cognitive processes are not the outcome of a parallelism between matter and mind—the latter emerging from the former, but as an event of the "intersection of mind and matter."[35]

We can extract a number of significant issues from Bergson's virtual image that might actually help us to better grasp the explanatory power of both the synaptic and proto-self within the interferences between neuroscience, philosophy, and art. To begin with, a Bergsonist conception of self would perceive the experiences that shape reality not as a *state of mind*, because the objects of experience are independent of mind. The mind does not, therefore, like the idealist would have it, shape matter. On the contrary, for Bergson, consciousness does not perceive matter. It is a memory, or more specifically, a duration *of* matter.

So, on one hand, although perception may seem to be *within* the self, or "truly within me,"[36] it is merely a contraction of "a single moment of . . . duration that which, taken in itself, spreads over an incalculable number of moments."[37] In short, perception makes the indivisible dividable.

Bergson's Brain-Imaging Experiments Expose the Illusion of Self

Bergson invents two highly effective protoneurological experiments by which to support his thesis. Indeed, if we were to begin by disconnecting our consciousness from the experience of matter, matter itself would become resolved into numberless vibrations. Matter would be a ceaseless continuum "traveling in every direction like shivers."[38] This is memory *persevered* in pure duration. Like Aldous Huxley on acid passing through the doors of perception, the self would free itself from these divisible spaces. The disconnected self "would obtain a vision of matter that is perhaps fatiguing for [the] imagination,"[39] but nevertheless, this would be a pure and stripped-out perception compared to the external perception we get before passing through one of Huxley's doors. When the trip is over, we must return through the door and reconnect with consciousness. We return to the divisible spaces with their "internal history of things," "quasi-instantaneous views"—a *pictorial* condensation of the "infinity of repetitions and elementary changes" of pure duration.[40] So the gap between representation and the thing is something like an artwork that tries to sketch out the thousands of successive positions of a person running in "one sole symbolic attitude."[41]

The second experiment asks us to imagine a kind of prophetic brain-imaging technology that is able to penetrate the gray matter so that we can observe the sketch work the artist produces and therefore capture pure duration. What would this artwork look like in the very matter of the brain? To begin with, Bergson contends that the brain state would indicate "only a very small part of the mental state,"[42] that is to say, the parts of the brain that translate pure duration into the artwork. But we would not see the representation itself. There is no whole photographic album in the brain! Any images of movement are emergent. There is no picture in consciousness "without some foreshadowing, in the form of a sketch or a tendency, of the movements."[43] So what would the operators of Bergson's brain-imaging technology actually observe? They may well observe the details of these sketches of movement somewhere in the matter of the brain, as a trace of some kind, but, Bergson contends, such flickers would reveal nothing of any journey between physiology and psychology realms. The brain image captures nothing of the association between matter and consciousness, any more than "we should know of a play from the comings and goings of the actors upon the stage."[44] The point is that the movement image is not stored in any observable way. Nonetheless, the model of storage, and predominantly the storage of information, is a problem that continues to haunt the cognitive neurosciences, particularly now, in an age when Bergson's brain-imaging experiments are a reality and when the location of such a storage system has yet to be found. How can we take seriously the computer–mind metaphor when there is no evidence of a memory bank? Where is the storage of psychodata in the matter of the brain? Where, in a nutshell, are the representations?

It is interesting to note here that Malabou contrasts the plasticity of the synaptic and proto-self with the *cybernetic frigidity* of the computer brain.[45] Unlike the computer brain, the plastic brain does not send and receive messages from on high. The plastic brain is instead understood in terms of its capacity to receive form and give form.[46] It has a tendency to transform and be transformed. To be sure, plasticity seems to contradict the static information metaphors of cognitive science. Nevertheless, it is argued that Damasio and LeDoux's implicit system does not manage to offer a distinct enough alternative to the computer mind. As Bennett and Hacker argue, Damasio and LeDoux do not entirely exorcise the engineering and information metaphors from their work. LeDoux's affectively charged events do not, as such, shake off the brain as storehouse metaphor. Similarly, Damasio's assertion that there is no emotional experience without visceral stirrings retains a casual nexus between stored representational images (categorized psychological situations) and somatic states.

In many ways, then, the affective turn LeDoux and Damasio introduce to the neurosciences represents only a small bridge extending out from cognitive sciences to an implicit noncognitive future. It is nonetheless a bridge worth crossing. Indeed, although Damasio's brain is littered with information metaphors like the signals, maps, and signs that are supposed to make up the circuitry between implicit and explicit systems, he also hints at a sketchlike patterning, which, to some extent, coincides with Bergson's artwork in between things and representations. Significantly, the proto-self does not, it would seem, produce representations, or at least the kind of photographic images cybernetic models assume to fill up the memory banks of the human information processor. Damasio's images are indeed *mental* patterns within a "structure built with the tokens of each of the sensory modalities," including visual, auditory, olfactory, gustatory, and somatosensory data.[47] Moreover, the proto-self is not grasped as a "storehouse of knowledge" but rather as a *reference point* that participates in the process of knowing.[48] This reference is not a visual comparison. It is a reference waiting for the nonconscious brain to answer questions that were never posed. Indeed, it is only when an answer is received that the conscious sense of the proto-self seems to emerge. But is this reference point not, to some degree, the same pictorial illusion of self that Bergson suggests we find ourselves occupying outside the doors of perception? Damasio can only answer that the autobiographical self is the *knower who knows* the answer to the questions never asked.[49]

Similarly, LeDoux's synaptic self still has one foot firmly stuck in the information circuitry of the cognitive paradigm. In the following, he describes the journey from matter (neuronal stimulus) to memory as a production of representations:

> Information about the external world comes into the brain through sensory systems that relay signals to the neocortex, where sensory representations of objects and events are created. Outputs of each of the neocortical sensory systems then converge in the rhinal cortical areas, also known as the parahippocampal region, which integrates information from the different sensory modalities before shipping it to the hippocampus proper.[50]

Yet, despite conflating sense making with information processing, LeDoux, like Damasio's multilevel emergence of self, suggests that below the surface of the wrinkly higher-level mental life of the neocortex, the synaptic self is composed of modulating neurotransmitters that produce relatively nonspecific effects. There are thus no *precise* representations of stimuli found in these deeper regions of the brain.[51] Instead, monoamines, for example, a class of modulators, including substances like serotonin, dopamine, epinephrine, and norepinephrine, can produce global state changes, spanning the valence of arousal necessary for fear, on one hand, and sleep, on the other, and distributing these states of arousal, via the many axons found in this region, to most parts of the brain, such as the converging zones of the neocortex and the hippocampus, where the *memory* of a concrete stimulus is assumed to become a mental representation, and where a perception becomes a conception independent of that original concrete stimulus.[52]

An assemblage theory of sense making needs to be able to provide nonrepresentational alternatives to the emergence of what Malabou problematically suggests is the semiotic nature of self-representation and mental storage. In short, emergent representation needs to be regarded as nonsymbolic and not explicitly stored.[53] That is to say, unlike the semiotic relation established between a symbol and meaning, nonsymbolic representations are a distributed "configuration of connection[s]," of excitatory or inhibitory neurotransmissions, which are connected to the world in a nonarbitrary way.[54] In other words, nonsymbolic representations are not static photographs but reproducible intensities linked together by association of, for example, colors and odors and the presence of food. This is because, and returning to Bergson's things that are independent of mind, we need to take into account an emergence resulting from a "direct accommodation or adaptation to the demands of an external reality."[55] The emergence of the intentionality of mental states, therefore, hypothetically, needs a body and a sensory environment to orient according to the stimuli of external reality.

The Politics of Plasticity

Where Malabou's espousal of the synaptic self is at its most effective is in its recognition of a significant lack of political direction in the neurosciences. To be sure, the synaptic self does not lead to a consciousness that knows itself. It does not, as such, escape Marx's problem of the false consciousness. Malabou's question is simply put: what kind of political project is arrived at where we eventually become conscious of our plastic brains? She notes that the plastic brain is the point of intersection where the biological and social meet.[56] We live in a reticular society, meaning that we are increasingly connected to each other through a thickening of networks, or meshworks, on which we are ever more dependent for survival. Of course, we could argue, following that famous and much older spat in the neurosciences, that these meshworks are in fact more synaptic, ergo rhizomatic, than they are reticular, as Cajal contended. Nonetheless, the issue to get across at this point is that within these novel connectivities between brains and environments, we find the point of exchange at which human biology and culture come together in an increasingly sociotechnical external reality. This is where, arguably, the emergence of neuronal intentionality becomes oriented according to the stimuli of sensory environments produced by market forces. We cannot consequently ignore the role capitalism plays in influencing the experience of this external reality and therefore influencing plasticity. This is what Malabou importantly draws attention to when she states that

> the guiding question of the present effort should thus be formu-
> lated: What should we do so that consciousness of the brain does
> not purely and simply coincide with the spirit of capitalism?[57]

The problem, it would seem, is that the neurocapitalist's embracing of plasticity has simply substituted the rigidity of a genetically deter-mined brain with a *flexibility* ripe for exploitation by market forces. Although Malabou is at pains to delineate the linguistic differences between plasticity and flexibility, it is important to stress here that such a distinction is not merely semantic. On one hand, plasticity is doubly defined semantically in the sense that it means to *receive form* and to *give form*. Malabou provides the examples of clay, in the first instance, and the plastic arts and plastic surgery, in the latter.[58] However, brain plasticity is also what we might refer to here as a *double event*. It has the capacity to affect and to be affected. When the plastic brain encounters a sensory environment, it can simultaneously be modified and modify its surroundings. Flexibility, on the other hand, is a very different kind of event. To be flexible means that something is supple and, as a result, can be easily shaped and adapted. So flexibility also *works on* and *or-ders* the thing it describes. To be flexible in the workplace, for example, means to be compliant to change or accept the worsening of conditions that make flexibility possible. As Malabou notes, this readily translates to flexible jobs and flexible factories.[59] Eventually, it translates to zero-hour contracts.

Flexibility is already well integrated into the cultural circuits of capitalism, but the circuitry is further enhanced when neuroscience is deployed in combination with efficiency analysis to effect change in the workplace, that is to say, using correlations between neural networks and behavioral patterns to effect change on workers who are, often for good reasons, like worsening conditions of employment, resistant to change. This is a major problem for the project to free the brain because, similar to the cybernetic frigidity of cognitive models of the mind, and its numerous reappropriations by the business enterprise, the freedom of plasticity is reduced to a flexible tool intended to put brains to work more efficiently as well as break down hard-fought-for cultures of resistance in the workplace. The ongoing symbiosis between the brain sciences and the establishment of *natural* laws of work and consumption, most notably apparent today in Damasio and LeDoux's appearance in business management and marketing literature, means that critical theory cannot afford to neglect the relation established between the brain and capitalism in which the freedom of plasticity is eclipsed by a compliant, obedient, and flexible workforce. To be sure, for Malabou, flexibility becomes the "ideological avatar" of plasticity.[60] It masks and diverts attention away from the affordances of freedom that plasticity might offer. In times of neurocapitalism, we are entirely ignorant of plasticity but not at all of flexibility.

Assemblage Brains Never Become One!

Following assemblage theory, plasticity needs to be grasped in accordance with the brain's relational encounter with sensory environments. This presents an alternative notion of how sense making emerges. From the embryo to the brain of a child, and the synaptic modulations and repairs to neuronal connectivity throughout life, the experiences of the sensory environments brains inhabit, and the access to material and expressive resources encountered there, are crucial to this emergence. The double event (the capacity to affect and be affected) means that brains are never simply the products of the society they are born into— as Durkheim's theory of collective consciousness contends. Indeed, like Deleuze and Guattari's ventures into the neurosciences, Malabou further grasps plasticity as the "refusal to submit to a model" of the brain, be it genetic, cybernetic frigidity or the flexible brain.[61] Instead, we find a brain that is not a ready-made. Plasticity is "an agency of disobedience" against all such prewired notions of constituted form.[62] It has been argued that the brain might have already achieved this freedom from the gene as neurons wrestled control away from the dominance of genetic code some six hundred million years ago.[63] Nevertheless, the important point to make at this juncture is that, unlike the synaptic self, the emergent plasticity of the assemblage brain never becomes a sum total; plasticity is always a *pattern changer.*

Throughout its lifetime, the assemblage brain becomes an open-ended work of art that has the capacity to continuously transform and be transformed by its sensory environment. There is no permanent synthesis of the self. Unlike Malabou, then, who counters Bergson's artwork by tracing the politics of plasticity to a dialectical movement that negates synchronic and diachronic forces to produce a synthesis of emergent identity,[64] the assemblage brain never becomes one. It is indeed curious that Bergson's creative evolution is drawn on by Malabou to explain the contradictory forces of energy that lead to the formation of the sum total of who we are. Bergson's notion of the explosion (bursts, discharges, and thresholds that transform nature into freedom) does not arguably mix well with a "dialectical play of the emergence and annihilation of form."[65] The multiple processes of subjectification seem to be a manifestation of a different kind of relational energy than that produced by the "primary and natural economy of contradiction."[66] Malabou's point is that there has to be a conflict between two kinds of energy: one seemingly *inside,* the other *outside.* On one hand, internal *maintenance* preserves the emergence of self, which needs to be synchronic—together at once and stable.[67] As Damasio argues, homeostasis refers to the "coordinated and largely automated physiological reactions required to maintain steady internal states in a living organism."[68] On the other, *change* affects this stability—change, that is, coming from events registered outside the brain that transform maintenance in creative capacities. The spiraling transition from the neuronal to the mental thus becomes grasped in this way as a translation of "neural maps" into "mental maps," taking the form of representational images (faces, melody, toothpaste, and memories of events).[69] This is a structuring of identity grasped in an emergent and stabilized self-fashioning, finely balanced between diachronic and synchronic forces; between the coat and the nail on which it hangs.

It is not that assemblage theory should discount oppositional forces altogether. It does meet with obstructions, but this does not mean that it encounters a dialectical synthesis of self-identity, a fine balance between destruction and formation, that produces emergent properties resulting in the totality of the living organism. On the contrary, it is possible to be both synchronic and diachronic without dialectical forces of negation leading to a sum total, that is to say, a diachronic and transversal emergence that does not generate a perpetual stability of emergent properties but is instead a relational plasticity with double capacities tending toward adaptations in response to interactions with sensory environments. So the apparent stability implicated in the sense of self might be understood not as a reified whole (or the representation of change) but as plasticity as change in itself.

The model of the synaptic self presents a mereological problem concerning the emergence of self-identity contingent on a contradictory tension between *inside* and *outside*. This is presented as a dialectical journey between the microlevel of the synapse and the macrolevel of mental stuff in which subjectivity is assumed to become whole—to become One. The ever-changing environment is awash with parts that push down on the microlevel of neuronal interactions while the self-maintaining subject constantly adapts so as to retain the durable stability of the mental state. Bergson would of course call such a dialectical flow between neurons and mental states an impossible, illusory movement. Matter is duration, and duration is a multiplicity. It divides up constantly. It changes in kind. It never becomes One![70] Identity is not therefore a formation or state; it is potential, a space of possibilities, a virtual surface.

The Phantom Thumb

Where in the mereological puzzle of the synaptic self are the other body parts? Where, for example, is Sholim's thumb brain? One way to approach this question is by referring to the notion of phantom limbs that have inspired the neurophilosopher Thomas Metzinger to claim that our sense of self is a necessary hallucinatory self-image.[71] That is to say, if the thumb is amputated, a ghostlike trace of it will appear in the brain's model of who we are. On the surface, this approach has some similarities to Bergson's illusory sense of self. As the phantom thumb scrolls across the screen, the smart phone user might sense this illusory trace as *being there* but frozen, unable to move. This haptic illusion can be further induced by way of using a mirror to provide a virtual thumb image, tricking the amputee into feeling the flicker of the scrolling movement. This illusory sense of self is necessary because, as Metzinger argues, the brain provides an internal image of a body, as a whole, so as to help an organism predict and take action according to the uncertainty of events it encounters in the environment. So accordingly, Sholim's severed thumb would need to know the world around it, its externality—even if it does not exist. The image of the thumb is located in the brain's systematic emulation of a whole body, which assumes an inner model of the outside body, even when that whole body is seemingly incomplete.

Unlike Bergson's illusory sense of self, Metzinger's phantom limb is grasped as a shadow on the wall of Plato's cave. Indeed, according to Metzinger, the cave is the brain itself. However, no one is living in the cave, just the shadows that are projected inward, giving the illusion that someone is dwelling there. This sense of self is nothing more than a shadow.[72] In short, the phantom thumb is explained by way of a locationist approach to the emergence of a sense of self, which does not sit well with Bergson's antilocationist stance, because it traces the phantom thumb directly to the cave brain. But is not the illusory Platonic cave a strange location in which to find a thumb? This inner model of the outside proves to be an unreliable shadow of its former sense of self, full of flaky representations. Thumbs and their phantoms surely inhabit a far more complex environment than a singular cave—a multiplicity? As Nietzsche puts it, "behind each cave [there is] another that opens still more deeply, and beyond each surface a subterranean world yet more vast, more strange. Richer still ... and under all foundations, under every ground, a subsoil still more profound."[73] Inside Plato's cave Nietzsche finds a labyrinth of caves, each with its own Bergsonian doorway to perception. So, in contrast to Metzinger's neuroscientific objectification of the body, in which the linear flow of messages seemingly travels through a cognitive ecology that never leaves the inside of the cave, Sholim's thumb might again be reimagined as endemic to a noncognitive ecology. This is not to say that the thumb does not think. But why do locationists assume that we sense its presence only in the brain? On the contrary, like Gramsci's muscular memory, the scrolling thumb (real or imagined) may indeed have a mind of its own that emerges as haptic intelligence. The illusory trace that appears in the cave brain may well have originated in the thumb itself, which, before being severed, forget to say farewell to the brain.

THE SYSTEMATIC BRAIN

How Wittgenstein's Thumb Learns to Think

The synaptic self is a locationist's brain. Although it is clearly linked to its environment through a nervous system that processes sensory stimulation, everything is controlled by the top-down influence of neuronal processing. More precisely, it is the neuron interaction itself that determines the sense of self (real or imagined). Nevertheless, following Bergson's antilocationism, it is possible to rethink the status of the thumb brain as conceived of as being in a certain place *inside.* An effective way to unravel such a puzzle is to perhaps follow an ongoing philosophical discussion within contemporary theoretical neuroscience that argues that by attributing thinking, as well as associated things like self-identity, to a material brain, in place of the immaterial mind, Cartesian structures are not so much circumvented as they are seen to reenter the synaptic self through the back door. The synaptic self marks, in this context, then, Descartes's guileful return. To be sure, as Bennett and Hacker argue, the causal relations established between the Cartesian mind and body are simply replaced by neuroscientific causal relations between brain and body.[74] In other words, simply by substituting the causal relations between the Cartesian mind and body with the causal relations between the material brain and the body, the cognitive neurosciences have retained the "overall conception of the relation of the 'inner' to the 'outer' that was enshrined in classical dualist thought."[75]

Descartes's return stems from something Bennett and Hacker call a mereological fallacy in contemporary neurosciences, that is to say, a deluded grasping of the whole–part relation that is assumed to bring about the emergence of consciousness.[76] To dispel this misconception, attention is drawn to two mereological paradigms that have influenced the brain sciences: Aristotelian and Cartesian.[77] The latter is not only evident in the nuances of contemporary neuroscience, they claim, but manifest in the historical foundations of brain science, as evidenced in the work of Sherrington, Eccles, and Penfield. Following instead the former Aristotelian model, Bennett and Hacker argue for a very different kind of theory of emerging consciousness, which, seemingly coinciding with Bergson, to some extent, challenges the locationist approach to representational memory storage. This Aristolean model is, however, traced to a significantly different kind of proto-brain-imaging experiment to the antilocationist approach Bergson encouraged. Indeed, this attempt to solve the puzzle of emergent delocalized consciousness was conceived of by Wittgenstein in 1953. Wittgenstein imagines a crude brain-imaging experiment consisting of a subject who is, at the same time, the experimenter, looking in at her own brain.[78] As a kind of future fMRI experiment, this allows the subject-experimenter to at once *think* and observe *thinking* directly in her own brain. In short, the subject-experimenter is observing a correlation of two things. On one hand, she sees *the thought*. This, Wittgenstein argues, "may consist of a train of images, organic sensations," or "a train of the various visual, tactual and muscular experiences which [s]he has in writing or speaking a sentence."[79] On the other, she oversees the experience of seeing her own brain work. Both, Wittgenstein concludes, can be regarded as *expressions of thought,* but the answer to the question "where is the thought itself?" becomes, accordingly, nonsensical. "It is misleading to talk of thinking as a mental activity," he says.[80] When a person thinks about writing, it is not the brain in isolation that thinks, but thinking is rather the activity performed by the hand. Likewise, when a person thinks of speaking, thinking is rather the activity of the mouth. Thinking is not therefore limited to the brain but also is the paper on which we write, the mouth with which we speak, and the thumb by which we scroll. This leads Bennett and Hacker, following Wittgenstein, to argue that to attribute thinking to the brain alone is a mereological fallacy, that is, the ascribing of attributes to parts that can only be ascribed intelligibly to wholes.[81] In other words, it is wrong to ascribe to the brain attributes that are ascribable to the human as a whole.

Bennett and Hacker's solution to the mereological puzzle is thusly rendered as a systematic mode of sense making in which it is the human who feels pain, not the brain. The brain is but a part. It does not feel sensations, perceive, or think. It is not conscious. The animal is conscious, not the brain. Brains do not have brains; they are brains. Humans have bodies; brains do not. Humans have minds; brains do not. A brain does not make up its own mind. A brain does not have a thought cross its mind. If we remove a brain from its body, they ask, and reconnect it to an artificial body, will a mind emerge that feels pain in this new body? Will the new body become thoughtful? Bennett and Hacker resolutely say no! Thinking is, they claim, attributed to the whole, not the part. A part does not act like a whole, so ascribing psychological attributes to a brain is a fallacy. Indeed, following Wittgenstein, it is the human who thinks, who sees, who is blind, who hears, who is deaf, conscious, unconscious, and feels sensations. In sharp contrast to Deleuze and Guattari's chaos brain, it is not the case that Wittgenstein's brain thinks. The human being (the whole person) is the system that thinks.

It is important to note that systematic sense making has two kinds of emergent properties that supervene the interacting parts of the brain. On one hand, there are *corporeal properties*. For example, a diseased or healthy body, and associated bodily sensations like pain or itching, are somatic properties. On the other hand, there are cognitive features *(incorporeal properties),* to think or remember, for example. But brains (and the other body parts) are not conscious, perceiving parts. Any relation established between the innerness of the incorporeal thought and the outerness of a body is an illusion. Such properties are only attributable to the human being as a whole, to the system. Indeed, although we seem to experience conscious thought in our heads as a predominantly visual representational space, this thought is only part of the illusion. There is no way, Wittgenstein argues, that thought is a visual space imprisoned in the brain.[82]

There are clearly useful contrasts to be made between the synaptic self and the systematic brain. These contrasts can indeed help us to productively grasp the interferences between contemporary neurosciences and philosophies of the brain. First, both models help us to move beyond the limitations of the cybernetic brain that has underpinned dominant theories of individual and collective consciousness in recent years. The synaptic brain does this by way of making a direct link between affective states and thought, as opposed to the cold calculations of the cognitive mind–computer metaphor. The systematic brain also seems to converge with Bergson's nonrepresentational movement image insofar as it too intervenes in purely representational systems of thought. Like Bergson, the antilocationist stance of the Wittgenstein-inspired systematic brain opens up further questions concerning the trend toward fMRI phrenology, that is to say, the problematic inclination toward correlating certain physical brain states, such as flows of blood and electricity, with mental states, including those located in the thinking and behaviors implicated in political belief, gender differences, and empathy, for example.

Both the synaptic and systematic models of self are arguably flawed though. While the synaptic self fails to completely exorcise the engineering metaphors of the cybernetic model, systematic sense making cannot similarly jettison the mereological fallacy from its explanation of emergent consciousness. Indeed, although it attempts to escape the inside and the outside, Wittgenstein's brain is still beholden to a sum total. It is, in short, a crude synchronic theory of emergence. The systematic brain returns us, as such, to the old problem of the One and the many. It is an emergence theory devoid of diachronic, transversal, and relational concerns, in which immutable wholes emerge with supervening properties that magically transcend all body parts. Even if we encounter Sholim's thumb in this system, this is not a thumb that thinks but an inseparable part committed to a gross generality of an emergent mind. Like the synaptic sum total of *who we are,* we arrive at yet another unrefined totality. As it turns out, the mereological fallacy (the part–whole problem) is the mereological approach itself in which the thumb is lost to the all-prevailing system.

So we return to Deleuze and Guattari's statement afresh. It is not the person or mind that thinks. It is the brain. So what kind of brain is this?

Detachable Thumbs

Before setting out the strange mereology of the assemblage brain, it is significant to remark on how the systematic model crucially misses out on the capacity of thumbs to become mentally detached from systems of collective sense making. Sholim's thumb is, like this, part of a system of brains, smart phones, telephony companies, and factory workers, but it never comes together to form a whole. It is not that a thumb's detachability makes it an unthinking part, though. It is a haptic sense maker. It thinks through sensation. The conformity of the scrolling thumb becomes, as such, more akin to Gramsci's muscular memory of the trade, a part of the rhythmic vibrations of physical work strangely detached from a collective consciousness.

There is also a crucial political detachment between the thumb and the worker's exploitation apparent in Sholim's artwork. This is a thumb that lacks empathy. It is blissfully oblivious to the labor used to manufacture the device on which it scrolls. There is, in the repetitive actions of Sholim's animated aesthetic figure, a ceaseless endeavor to scroll over (to overlook) the video of the Chinese worker. The GIF artist makes it difficult to escape the looping sensation of this concept, but perhaps, in *real* "命," most thumbs would scroll past this kind of content to peruse something more seductive. Indeed, the coming together of thumbs and mobile devices seems to provide a convenient noncognitive ecology—a collective nonconsciousness.

ASSEMBLAGE BRAINS

In light of the assertion made by the followers of Wittgenstein, that is to say, that it is the person (mouth, ears, eyes, brain, thumbs, etc.) who thinks and *not* the brain in isolation, we can now clearly distinguish between the systematic brain and the chaos brain. This is because it is Deleuze and Guattari's brain, not a mind or a person, that thinks.[83] To be sure, there seems to be very little by way of a productive interference here. Wittgenstein turns out to be the straw man of Deleuzian ontology: a philosophical black hole, the *assassin of philosophy.* Worse still, the Wittgensteinians—these masters of philosophical "catastrophe"— have, by producing a hybrid of logic and phenomenology, imposed a system of absolute terror! *"One must remain very vigilant."*[84] As a result, assemblage brains are elaborated on here through three stages, each evidencing, to some extent, how Deleuze's suspicions concerning Wittgenstein become interwoven into assemblage theory. First, a Proust-inspired rethinking of the whole–part relation, and borrowed from *Anti-Oedipus,* is deployed. Drawing on Proust enables Deleuze and Guattari to describe how the parts of an assemblage become detached from the forces that endeavor to bring parts together into sum totals. Clearly this is an intervention into the systematic forces that bring Wittgenstein's thumb into relation with the person who thinks as a *whole.* Then, second, I follow a trajectory to more recent articulations of assemblage theory, which similarly provide an alternative to the synthetic unifying forces of the One and the many. Again, parts become detachable from the forces that endeavor to unify them. Like this, body parts are not simply rendered as a property of a thinking system but can, like Sholim's thumb, and Gramsci's muscular memory, become detached from cognitive ecologies to form novel assemblages of sensation. This is an important move, because by eluding the generalities of an emergent mind system altogether, assemblage theory draws attention to the capacities and tendencies of these detachable parts, which are not grasped as a whole thinking subject or collectives of persons but as an alternative protosubjectivity in the making. Third, I return to a Tarde's monadological endeavor to undo the whole–part problem, beginning with his rejection of Durkheim's collective Ego, before moving on to think through how Tarde's appeal to panpsychism allowed him to bypass emergent consciousness and imagine every entity as possessing some form of monadic subjectivity—a population of microbrains.

A Proustian Puzzle

To begin with, and following the early development of assemblage theory (or the desiring machine) in *Anti-Oedipus,* body parts need to be considered as a Proustian puzzle, that is, a strange rethinking of the mereological problem. Instead of regarding the interaction between parts as limited to the emergence of wholes, assemblage brains are (1) referenced to the aesthetic figure of memory Proust introduces in *In Search of Lost Time* and (2) traced to an understanding of brain–somatic relations exposed to the force of encounter with the event. Importantly, this encounter does not result in a complete solution to a puzzle. A memory, for example, is not something that is pieced together to form a complete recollection of events. Assemblage brains provide a strange approach to mereological thinking insofar as they to do away with a notion of memory function constituted in a unifying system.

Although alluding to an antilocationist theory of emergent subjectivity, the problem with the systematic brain is that it does not consider the force of events in which the parts of a memory become lost to the events of memory. As Proust sets out, these parts can be swept away by a gust of wind. They are always in passage; they can be wasted and lost in the moment. The images of memory, which Bergson similarly considered to be merely a divisible version of the undividable, are not stored in a readily accessible memory bank containing the objects of experience but have to be persistently regained by way of an encounter with events. Memory is not therefore an inert solution to a fixed puzzle but is constantly thrown up in the air, requiring the person who remembers to capture just a part of the becoming image in a truly schizoid artwork. In assemblages of sense making, the parts of memory are not therefore coupled to supervening wholes. They are not reduced to dependent properties but instead become defiantly detachable. The interaction between parts does not, as such, become beholden to synchronic emergence but is subject to synchronic, diachronic, and transversal processes of becoming. Caught in the event of a memory of a former lover, for example, the Proustian parts of memory can suddenly, and involuntarily, emerge, triggered by a sound, a smell, a taste, an image. Indeed, these parts of memory can also become associated (or disassociated) with variable feelings: fear, joy, jealously, love, or hate, perhaps, depending on how the memory is re-membered.[85] Every memory is dependent on relationality of some kind. External objects play a role, too. As Ellis and Tucker put it, a past memorable event, like a childhood birthday, is "enacted through the re-engagement with objects associated with that time."[86] This is not a pure act of cognitive retrieval from *inside one's mind,* as cases of posttraumatic stress disorder, triggered by, or associated with, a relation to certain objects or events, seem to suggest.[87] Memory recollection is not, therefore, reduced to a cognitive function (location or system) but to a *process* Schillmeier well describes as "re-assembling the lived but veiled affective relations, their members, detached attachments and practices that are configuring... relations of the world."[88] In short, Proustian memory is noncognitive.[89]

To consider memory as an assemblage, we need to throw all the pieces of Wittgenstein's puzzle in the air and see where they land. But this does not create another puzzle. These parts will be thrown in the air many times again. Indeed, the parts of a memory-event do not belong to any one puzzle but to many! This is where Wittgenstein slips up. His puzzle endeavors to be whole—to bring together the functions of the brain, like memory, into a unified system that thinks—but wholes will never become the sum total. The parts of a locationist and systematic brain would need to be assembled by "forcing them into a certain place [or system] where they may or may not belong," their "unmatched edges violently bent out of shape, forcibly made to fit together, to interlock, with a number of pieces always left over."[90] Memories are not the result of strong emergence, as such. Again, the schizoid mixture of synchronic, diachronic, and transversal emergent forces does not come into being but remains an open-ended process of becoming:

> In desiring-machines everything functions at the same time, but amid hiatuses and ruptures, breakdowns and failures, stalling and short circuits, distances and fragmentations.[91]

All this happens "within a sum that never succeeds in bringing its various parts together so as to form a whole."[92] So although an assemblage may appear as a totality or a unity alongside particular parts, it does not totalize or unify these parts. This is why parts are Proustian; they are subjected to the flow of time, making the emergence of memory pass through asymmetrical sections or paths that suddenly come to an end. A memory that is regained needs to be prized out of "hermetically sealed boxes, non-communicating vessels, [and] watertight compartments."[93]Although we might think of memories as contiguous, they have *gaps* in a fragmented cogito. They are pieces of a puzzle that never quite fit together.

More thought needs to be given to the interaction between parts. What brings assemblages of re-membering together, for example, is not the sum of parts equal to the emergent global persons, integrated wholes, or systems of subjectivity but rather are the relations established between component parts that are detachable. If we are to consider the assemblage as a puzzle, then it is one full of partial parts, never waiting around long enough to become a complete object. Once more, parts are not the missing pieces of Wittgenstein's unfinished puzzle. They are orphan parts or Proustian memories without parents. So we do not need to locate all the pieces in the brain in one place or in a system to make up the whole. As Bergson argues, conscious processes, such as those experienced in recollecting the past, are not preserved in the brain. They are preserved in themselves—in their duration, in matter. There is no mind behind the brain.

Avoiding Generalities of Mind

Complex emergence theory suggests that we do not need necessarily to follow the assemblage brain through to an emergent mind. Irreducible wholes, like the mind, should not be considered legitimate inhabitants of objective reality because they are nothing but reified generalities.[94] We should not waste time specifying a mechanism of emergence for a mind in general. Better to account for the emergent properties, capacities, and tendencies of such concrete wholes by looking for their relational assemblages. In other words, subjective consciousness need not be explained by way of an emergent mindful self-identity but should be grasped instead as a multitude of outcomes resulting from encounters with affective environmental stimuli that trigger variable responses to subjective gradients. Consciousness is not therefore understood as a systematic outcome of becoming a person (a self-identity) but becomes a series of reverberations following on from a series of visceral responses to the environment, ranging from fundamental habituation of movement (e.g., avoidance) to expressive physical and emotional feedback to necessities like nutrition, producing hunger and thirst, and reactions to pain, like anger.[95]

To confront Wittgenstein's systematic sense of self, we need to grasp a strange relation between parts and wholes in which the former can become detached from latter. In short, parts can become detached from assemblages of relation and still live on. This, of course, contradicts Bennett and Hacker's example of the failure of the neurosciences to thus far carry out a successful brain transplant. They claim that even if such a transplant were possible, we should not expect to find that the sense of self of the donor emerges in the recipient. But these brain transplant failures might be nothing more than a prolonged moment of temporary epistemological ignorance. We can, as such, point to other significant body parts that we now know to be detachable. The heart, for example, can be transplanted and resume regular functionality in another body.[96] Thumbs, too! Indeed, this is more of an ontological question than an epistemological one. The parts of an assemblage are, like Proustian parts, orphans that can always be adopted. To be sure, parts can function as parts that synchronously bring together territorial interrelations, but at the same time, their detachability, irreducibility, and decomposability mean that they can oscillate between deterritorialized and reterritorialized interrelations over varying time periods.

Nonetheless, although the parts of an assemblage can be considered as independent from these territorializing forces, there is no escaping, it would seem, the sense that many people have of subjectivity as a whole. The question remains as to why self-identity seems to be experienced as a sum total of who we are. How can this be, when subjectivity is (1) becoming, (2) fragmented, and (3) not stored anyplace? Beyond Bergson's doors of perception, assemblage theory offers two possible explanations for what we might consider to be an illusion of the totality of a coherent self.[97] To begin with, and significantly at odds with the synaptic self, the assemblage brain is not reduced to neurons or their physical synaptic connections. These are the relations of interiority that essentialists lean on to explain their quidditous sense of themselves as a territory. Importantly, an assemblage theory of protosubjectivity presents us with a relation of exteriority. It is a highly deterritorialized becoming, as such. Nothing is symbolic or stored in locations. There is only a possible configuration of connected strengths and weights produced by environmental stimuli. There is, like this, no relation established between inside (neuron) and outside (environment), just a radical relationality that undoes the relation between the two (see chapter 5). There is no emergent system moving from micro- to macro-levels. Everything is already on the outside. *Everything is environmental.* Indeed, environmental stimulation produces the capacity to create a protosubjectivity. However, deprived of stimulation, these prototypes are just a potential of an unexercised capacity: a space structured only by possibilities introduced by exposure to the outside.

Second, subjective identity may be explained as an effect of emergent habitual and routine behaviors. These are "habits [that] territorialize behavior making certain patterns relatively uniform and repetitive."[98] So the assemblage brain is not a symbolic information storage system but rather a process whereby deterritorialized prototypes may become exposed to habitual behaviors, making protosubjectivity repetitious, and subsequently seeming to be coherent. There are evidently cultural and political dimensions to this second aspect of the illusion of self. Indeed, unlike Metzinger's illusory ego, which seems to have emerged from evolutionary necessity alone, the repetitions of Sholim's thumb brain draw attention to subjectivity in the making responding to the sensory environments of neurocapitalism.

Significant conceptual tools can be drawn from assemblage theory to help understand how what appears to be an illusory sense of wholeness is the outcome of continuous open-ended emergent processes. First, in lieu of strong or weak emergence, wherein transformations occur all at once, protosubjectivity needs to be understood as a dynamic historical process of territorialization, deterritorialization, and reterritorialization. The brain is not an essential unchanging entity. It is, like every other part, caught up in these historical processes that compose the sense of self. Second, diachronic emergence helps us move beyond an explanation of emergence theory requiring the macro- and microinteractions, such as downward and upward causation, in favor of *modular agents* that can pass through all intermediate scales.[99] What might appear to be systematic unity at any one of these levels might be expounded by a tendency for the swarming of the society of events to give the impression of wholeness.

Third, the notion that simple emergent properties are able to somehow transform themselves into complex wholes is reversed in assemblage theory insofar as it is the complex relation between parts (including the assemblage of properties, but also potential capacities and transformative tendencies) that produces seemingly coherent events like those that might be experienced as a sense of self. Indeed, the properties of parts, that is, their quiddity, their resemblances to each other, their *whatness*, and so on, are never a given and should not therefore be relied on to guarantee or sustain the illusory totality of a whole. Parts are contingent on affective relations to other parts, for example, their capacity to come together or fall apart. A society of events, like those that apparently maintain this stable sense of self and make memories seem whole, only needs to reach a certain threshold point before the events are transformed into altogether different states. So returning to Proust, we might grasp such a threshold at the point in which the difference between time lost and time regained becomes insurmountable. Such is the case when, as Schillmeier contends, the event of dementia "troubles the artful process of regaining time and its repetitions whenever time is irretrievably lost," producing an "abyss of not-being."[100]

Tarde's Monadology (a Slight Return)

The intersection between assemblage theory and Tarde's monadology is a good point, I think, at which to return to the problematic theories of the emergence of the consciousness of the consciousness apparent in Durkheim's social theory and the megaminds of 1990s cyberculture. Indeed, if we are to question the logic of supervening sum totals, and systematic wholes, we need to account for alternative explanations of how mind and matter are held together. Like much of Tarde's work, *Monadologie et Sociologie* (1893) can be read as a refutation of Durkheim's dynamic collective consciousness and, as such, as an early rallying cry against the related claims of strong emergence. Published three years earlier, Tarde's imitation thesis had already countered collective consciousness by focusing on the capricious emergence of mostly non-conscious associations between parts (individuals) brought together by forces of imitation, influence, and magnetic fascinations. These mostly capricious associations become social; indeed, everything (organic and nonorganic) is determined by a universal sociology, but not in the sense that they become stable social wholes.[101]

In *Monadologie et Sociologie*, Tarde goes further in his endeavor to reverse the one-way flow from parts to wholes by avoiding systematic relations between insides and outsides, or upward and downward causation, and, instead, drills down into the infinitely small. By doing so, he uses the monad to substitute the emergent mind with a merging of matter and mind into a single monad, which thereafter proliferates into the dust of the pure void.[102] Yes, there are things that come together, things that are *possessed* by vital forces, but beyond these emerging territories, there is only the infinite dissolution of matter into the abyss. Indeed, Tarde finds it very strange that a collective consciousness can be derived from "*relations between distinct beings* [which] *can of themselves become new beings numerically added to the former.*"[103] That is to say, he questions a logic that starts with a prototype of collective being at an atomic microlevel of creation before being transformed into a transcendent higher level or social whole, as in such circumstances when the interaction between atoms results in the emergence of an ego. In such cases, we are forced to admit, Tarde contends, that a

> determinate number of mechanical elements enter into a certain kind of mechanical relation, a new living thing which previously did not exist suddenly exists and is added to their number.[104]

More specifically with regard to the emergence of mind over matter, Tarde draws attention to an unrealistic model of consciousness in which "a given number of living elements find themselves drawn together in the desired fashion within a skull," resulting in

> something as real . . . , if not more real than these elements . . . created in their midst, simply in virtue of this drawing together, as if a number could be increased by the disposition and rearrangement of its units.[105]

From this point on, there would be no reason to stop at an emergent mind, Tarde argues, so that every *harmonious* relation between elements produces a "superior element, which in turn assists in the creation of another yet higher element," and so on and so forth.[106] Indeed, the persistence of this unidirectional ratcheting up through the micro- and macrolevels of creation can, if it is not constrained, be traced not only all the way from the atom to the ego but beyond, to the impossible emergence of what Tarde calls a *collective ego*.[107] Such a thing, he argues, cannot be considered as anything more real than a marvelous metaphorical relation.

Comparable to assemblage theory, Tarde's monadology does not begin with simple properties to explain complex phenomena. In contrast to understanding individuals as "indivisible atoms, homogenous and admitting of movement only as a whole," he looks to the infinitesimal and highly complex constructions, and architectures, animated by *highly varied internal movements*.[108] A monadological understanding of collective consciousness would, in this way, take into consideration one member who might personify a whole group and "individualize it no less entirely in themselves."[109] This might be the case when the magnetism of a charismatic leader emerges from a group. But this personification is not *collectively born*. The leader is not *One* of us! The leader is always a member of the group yielded from the interactions between other individual elements. He is therefore nothing more than an individual, as is the group, or the city, or the nation he, and the group, derive from; all are individuals, all are parts. These are the many! They are a Tardean multiplicity that never becomes One! The global leader is nothing more than an accumulation of *facts* about leadership that vibrates through an assemblage of associated brains. Indeed, the monad, when left to its own devices, achieves nothing more than this *tendency to assemble* and resonate. So a Tardean collective encounter is not the production of supervening wholes but a mostly unconscious relation between brains being directed by what Tarde considered to be the mostly accidental forces of monads.

There are nonetheless varying lessons to be learned from Tarde's monadology, some of which help us to grasp the potential freedom of assemblages of sense making, and others that might limit its agency. Indeed, although mind and matter are folded together in Tarde's monadological cosmos, it is important to note that the concept of mind, or indeed subjectivity, is not entirely discounted from the monadological cosmos. This is not to say that the Cartesian split between subjects *with* and material objects (or animals) *without* a mind is retained. As Theo Lorenc points out, in Tarde's theory of panpsychism, "every entity has some form of mind, self, or subjectivity."[110] The point is that there seems to be a capacity for a universal mind that is not exhausted by conscious states or cognitive processes alone. Similar to Leibniz, whose percepts existed below the threshold of consciousness, Tarde finds the possessive powers of belief and desire fundamental to a wider concept of sense making. In short, the mindlike qualities of organic and inorganic matter are similarly driven by forces of desire and a belief in their own consistency. The possession of matter—what brings things together—is, in other words, a double emergence of collective nonconscious desire and a cognizant belief in processes of assembly, that is to say, the power of universal social relationality.

There is, however, a seemingly perturbing lack of democracy and reciprocity apparent in the possessive forces that bring things together in the monadic cosmos. In short, Tarde seems to offer us degrees of monadic perfection in which the mind-matter of the human being sits atop of lesser inorganic mind-objects. As Lorenc argues, although there is no "deep qualitative chasm" established between humans and nonhumans, there is a marked "consequential quantitative difference" apparent in *Monadologie et Sociologie*.[111] This is because the human appears to have a possessive advantage over the inorganic world, because she has the capacity to hold the images of objects in her mind, to possess them in a representational space and, thus, bring them within her field of action. Cognitive representations therefore equate to some kind of special place in the monadic dominion. However, this interpretation of Tarde might well confuse possessive agency with what Didier Debaise calls "the action of 'taking possession' of an object by subject."[112] In other words, by giving more power to the sense of belief (understood as cognitive and representational) than to desire, the *dynamics of power* in the monadic cosmos are reduced to simplistic oppositional power relationships.[113] Instead, the possessive forces of Tarde's universal sociology need to be considered as more subtly distributed by way of immaterial flows of influence, imitation, attraction, magnetism, hypnosis, sleepwalking, and so on.[114]

What brings things together in a Tardean relational encounter is not tilted toward either belief or desire but occurs in the intersection between the two. It is the inseparable *in-betweenness* of desire and belief that counts: in-between the thing and the idea, representations and affective states, intentions and passions, concepts and sensations. Indeed, quantitatively speaking, we might even consider a shifting dynamic of power in which what the inorganic world lacks in self-belief it makes up for with the forces of desire, and similarly, the organism's capacity to comprehend itself, and the inorganic world it encounters, in a representational space is matched by a deficit in its grasping of the forces of desire that bring about its social becoming. To be sure, Deleuze and Guattari grasp Tarde's theory of social emergence congruently as nothing to do with the movement from microlevel individuals and the macrolevel of the social; rather, it is between

> the molar domain of representation, being collective or individual, and the molecular domain ... where the distinction between social and individual loses all meaning.[115]

Perhaps the most fruitful domain to consider when pondering the emergence of a proto-self is not the molecular or molar in isolation from the other, or indeed the journey from one to the other, but instead the intermediary relations that assemble organic and inorganic worlds. Certainly protosubjectivity is not to be found in the molecular or molar worlds alone but traced through the transitional arrangements that traverse these two worlds.

Monads and Microbrains

Returning to the conception of the megamind, it must be noted that the desire for collective modes of sense making should not be dismissed outright. It is just that we need to rethink the narrowness of the early appeals to purely cognitive ecologies expressed in the cybernetic models. It must have seemed plausible to those stuck in the cognitive paradigm to assume that the main function of a brain was to produce concepts, and for a collective consciousness to come together, these concepts would need, perhaps, to spread out to form a global tissue of ideas. But as the emotional brain thesis argues, the production of concepts and emotions is enmeshed in synaptic networks, and this enmeshment does not end in one brain but expands outward into empathic relations, for example, socially communicated through electronically mediated sensory environments. It is indeed the propensity to assemble and resonate in more broadly conceived environments of sense making that becomes the most significant feature of the assemblage brain. This is a sense making produced in the force of encounter with events (in the Whitehead sense) and transmitted throughout the monadic cosmos.

To fully grasp these monadic propensities in sense-making encounters, we need to return to *What Is Philosophy?*, first to revisit the strange discreteness established between concepts and sensations. To be sure, Deleuze and Guattari seem to retain their position as philosophers of mixture insofar as they see the chaos brain as no less sensation than it is concept. Both concepts and sensations are thus relational to the forces of events. They are preservations and contractions of events. There is, however, a clear distinction made between the two. While the conceptual brain of the philosopher is seen as a counterpart of the capacity to affect and to be affected by encounters with events, concepts are nevertheless limited to their response to placements, movements, changes, ordering, and relations established with other concepts. In contrast, sensations preserve excitations and vibrations from environmental stimuli. Perhaps the philosopher's concepts can only ever really provide an idea of affect, and the artist's sensation only submit to an affect of concept, but sensations are already in the world, bringing body parts into relation with each other. Sensations might therefore have a more subtle encounter with events in the assemblage brain.

Second, there is a need to distinguish between phenomenological relationality and a kind of relationality that is not bound to human subjectivity but can be extended to nonhuman and inorganic worlds of microbrains. These microbrains can also, it seems, thrive as sense makers. They do not, however, constitute a megamind. We need to ask what kind of relational encounter this is. Although assemblage brains are contingent on relations to the world, it is important to note that this is not relationality grasped in the phenomenological sense of being in the world. The relationality of sense making is continuously being made in its confrontational encounter with chaos—a limitless encounter with the world understood through multiple processes of subjectification. Indeed, this is not a subjective production of mental objects or experiences of events. There is no subjective point of view, looking outside from its inner self beyond that of a monadic protosubjectivity, that is to say, a kind of paradoxical crossover between a monism, which declares that *"matter is mind, nothing else,"* and a pluralist domain of multiplicity, which does not surrender itself to a "kind of subjective idealism that would state that matter is only representation or idea."[116] As Debaise well describes a Tardean sense of monadic subjectivity, "matter is [not] a product of the mind, but that it is already, so to say, mind from the inside." This is significantly a representation not of the world but of a universe composed of others whose beliefs and desires are related. Indeed, this "mind from the inside" would seem to conform to Deleuze and Guattari's brain that thinks and, in turn, resemble how Ruyer defined the brain's encounter with the world, as a *form in itself* that does not "refer to any external point of view, any more than the retina or striated area of the cortex refers to another retina or striated area of the cortex."[117] So assemblage brains should perhaps be grasped through the affective flows of noncognitive desire as well as cognitive ecologies of self-belief. The latter, though, are importantly grasped as nonrepresentational and nonsystematic antilocations, the former raising interesting questions concerning what Robert Zajonc describes as affect's capacity to think for itself; that is to ask, what would it be like if the sensations of affect had a mind of their own? Sensation may, like this, be considered a kind of sense making that is perhaps closer to protosubjectivity.

CONCLUSION: TO THINK THE PROTOTYPE IS TO THINK THE UNTHINKABLE

There is much to learn about sense making by looking for potential interferences established between the scientist's functions, the philosopher's concepts, and the artists who live in the world of sensation. Indeed, the significance of the aesthetic of Sholim's 命 is that it goes some way in helping the philosopher think like a thumb. Perhaps this revelation should not be so surprising. It is after all the work of the artist to think in sensations (or the sensation of concept). By doing so, the artist hooks up the audience to a mostly noncognitive ecology. The problem for philosophers, who are used to dealing only in concepts, or in concepts of sensation, is how to think, talk, and write in the medium of the pure sensation. This requires a nonthinking philosophy, or perhaps what Deleuze and Guattari call *nonphilosophy,* that is to say, an interference that submerges the distinctiveness of concepts and sensations (and functions) into the chaos the brain plunges into.[118] There are those philosophers nevertheless who seem content to remain stranded in a cognitive ecology where they continue to engage in uncovering hidden ideas and turning them into the subjective reconstruction of objects, aka representational thinking. The assemblage brain must not, in contrast, become trapped in the cavernous cranium (the cave brain) for fear that it might never come to sense protosubjectivity. But then again, to even begin to sense the assemblage brain, we will require a kind of sense making that cannot be thought. To be exact, to *think the prototype* is to think the unthinkable.

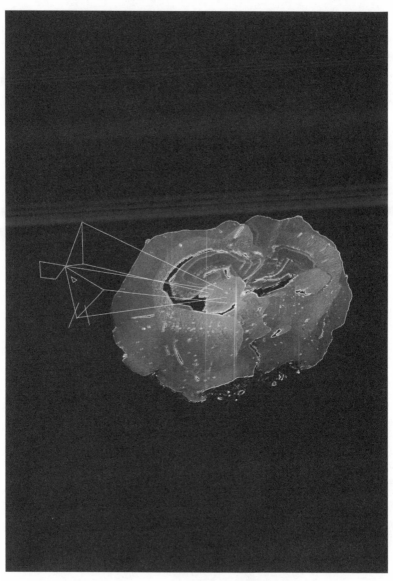

The lab rat awakens. Illustration by Francesco Tacchini.

5

Relationality, Care, and the Rhythmic Brain

THE PROBLEM OF THE INSIDE LOOKING OUT

Despite the tendency for Deleuzian ontology to lean heavily on the brain sciences, there seems to be a considerable mismatch between this commitment and the assemblage. Unlike the radical relationality of assemblage theory, the neurosciences tend to present a locationist theory of subjectivity constrained to an *inside–outside* relation. Like this, the *essence* of who we are on the outside is all too often grasped as determined *in* the synapses. This is a locationist theory of subjectivity very much stirred into action by recent brain-imaging technology but that can perhaps be traced back historically to Cajal's anticipation that the neurosciences would, one day, discover the *butterflies of the soul* deep inside the complex networks of neurons he observed. Similarly, the behaviorist's black box, although responding to and shaped by external stimuli, is nevertheless a model of the outside retained on the inside. The inner workings of Pavlovian circuitry, which Cajal hypothetically located in the cellular occurrence, were, subsequently, conceived of as what governs conscious mental phenomena. The shift from the behaviorist to the cognitive paradigm changed very little. Indeed, although the cognitive sciences promised to open up the black box to the outside, revealing the information circuitry that related the computer mind to the external world, the engineering metaphor simply added further levels of separation. The cognitivists problematically partitioned the brain and the mind into hardware and software processes, wherein the software, stored inside the hardware, contains a representational model of the outside world. Moreover, though, as the brain has been opened up to imaging and physiological experimentation, a complex lower level, comprising a hundred billion black boxes, is now assumed to contain psychoneural representations of the outside.

Although claiming to undo Descartes's substance dualism, in many ways, Cartesian structures remain intact in many of the theories forwarded in the neurosciences. Indeed, although materialist accounts claim to have moved on from Sherrington and Eccles's neuronal Cartesian theater, they have, it would seem, merely substituted the old dualism that makes the mind an immaterial substance for a new duality of causal relations between a deterministic brain, on one hand, and the behaviors, psychologies, and emotional responses to the world, on the other.[1] The conception of the relation of the inner to the outer "enshrined in the classical dualist thought soldiers on more or less intact."[2] Furthermore, albeit incorporating affective relations, ergo demonstrating a brain's capacity to affect and be affected by an environment, the plasticity of LeDoux's synaptic brain, and Damasio's protosubject, falls back on the same engineering metaphors. The emotional turn in the brain sciences conceives of emergent subjectivity as something that is stored as a sum total of who *we* are in the world, stowed away someplace in the interior of the synapses. The problem is that the significant role of politically constituted sensory environments in processes of multiple subjectification, like those produced by market forces, for example, becomes lost in a general tendency to locate subjectivity in the making as something that needs to be approached from *within*.

This concluding chapter approaches a politics of sense making understood in relation to the outside forces of sensory environments. It does this, initially, by contrasting assemblage brains with Metzinger's fairly recent evocation of the engineering metaphor of virtual reality to describe what he see as an evolutionarily constituted illusory sense of self.[3] In short, on one hand, Metzinger persists with an inner model of the self, in which the essence of conscious subjective experience is conceived of in a strange solipsistic rendering of the external world. Indeed, Metzinger's model of the virtual self introduces a further level of duality between an illusory dream state and an objective reality too complex to contemplate. The only way to free the brain from this false sense of self, Metzinger argues, is to alter the state of consciousness so that *we* are able to *see through* the illusion of the sense of self. The assemblage brain, on the other hand, puts the external relationality established between parts ahead of the emergence of inner models of the whole body (illusory or otherwise). This is an approach supported in this chapter by returning to Tarde's social brain and evoking Deleuze's similar appeal to the force of the outside, wherein the inside becomes nothing more than a *fold* of the outside.[4] So as to open up the political dimensions of the fold, I also draw on John Protevi's account of radical relationality (a reading of Bruce E. Wexler's brain and culture thesis),[5] while also expanding on Christian Borch's rhythmanalysis of imitative relations.[6]

The final part of the chapter engages with a notion of the rhythmic brain's relation to outside forces by returning to the problem of attention deficit addressed in chapter 3, in particular, the development of care systems intended to manage the problem of inattention. Here we see how the neurosciences continue to render ADHD an anomalous condition of the internal brain, wherein the genetically derived malfunctioning of dopamine neurotransmission, for example, determines the inattentive subject's anomalous encounters with the outside world. My focus here is, for that reason, not on the deficit itself but, following Stiegler, on the politically constituted environments of popular industrialism and care systems in which the problem of attention persists. Of course, ADHD has been widely critiqued as a socially constructed and culturally specific pathologization of inattention (and troublesome hyperactive youth behavior), but little focus has been applied to the demanding sensory environments in which inattentive states are composed, cared for, and perhaps even celebrated. It is indeed in the systems of carelessness of neurocapitalism and the marketing of neuropharmaceutical remedies for attention deficit that we find the kind of dystopian control of nonconformity we encountered in the first half of this book. Like this, *care* is realized in the overprescription of psychostimulants and EEG-driven brain frequency entrainments that sustain the requisite and neurotypical speeds necessary for life in the sensory environments of the marketplace and the schoolroom. The problem, it would seem, is that these environments (and modes of care) are constituents of a power system that not only targets the inner brain in isolation but also manages brain frequencies according to the force of the outside. Consequently, the book concludes with a focus, not on a notion of brain freedom that is realized in the neuronal dream states of an inner world, but rather on a brain–somatic relation that only knows an outside.

TUNNELS AND FOLDS

Given its tendency toward solipsism, it is intriguing that Metzinger makes use of the mirror neuron hypothesis to support his concept of the *Ego Tunnel*.[7] Indeed, at first glance, there are some interesting parallels between the passivity of the Tardean brain's dream of action (as encountered in the somnambulist) and a tendency in the neurosciences to conceive of a brain that dreams itself. For example, the porosity of Tarde's brain to the vital forces of imitation repetition is seemingly echoed in Metzinger's neuroscientifically inspired vision of subjectivity "swimming in an unconscious sea of intercorporality."[8] This is a subjectivity that is similarly made, it appears, in the perpetual imitation of self and other to a point where the two might become indistinguishable. But if we look again, Metzinger's flirtation with the mirror neuron presents a poorly defined relation to the outside. To be sure, he follows a brand of social neuroscience that locates the social *in* the self-model; that is to say, the subjective feeling of the other "pops up in the Ego Tunnel" as a representation of another Ego Tunnel.[9] Metzinger's appeal to the imitative capacity of the mirror neuron is, like this, arrived at through a *dream tunnel,* which, despite encountering other tunnels, always leads back to a model of selfhood projected onto the walls of a sealed cave. Subjectivity begins in this cave, as a concept of itself. The encounter with the other is *grasped,* similarly, as a matching concept—a mental model of the other, taken *in* and stored for future reference.[10] For this reason, there needs to be a clear distinction made here between the conceptual tunnels of Metzinger's cave brain, in which the ineffableness of sensation always needs to be located (and revealed) in the neural correlates of the cave brain, and the folded forces of the outside that constitute the potential of affective encounters with the Tardean brain.

INSIDE METZINGER'S CAVE BRAIN

Despite its claim to be a neuroscientifically supported materialism, it is not the brain but the virtual model that thinks in Metzinger's cave. This is a psychoneural theory of consciousness in which the inner workings of the brain provide a dreamlike representational model of the outside.

Primarily, the Ego Tunnel extends the engineering metaphors of cognitive science to a "special part of... virtual reality" in which we find a software interface that conveys information about the actual world so that a mental model of that world, and the user's place in it, can be stored away for future reference. As Metzinger puts it,

> by generating an internal image of the organism as a whole, it allows the organism to appropriate its own hardware. It is evolution's answer to the need for explaining one's inner and outer actions to oneself, predicting one's behaviour, and monitoring critical system properties.[11]

There are, however, a number of problems with this recourse to an evolutionary constituted self. First, despite its appeal to imitative capacities, the Ego Tunnel is a locationist theory of subjectivity. What is "phenomenally represented in a mind" inhabits a "volume of space."[12] Certainly, for Metzinger, conscious experience is a strictly "internal affair." It is "something very local, something in the brain itself,"[13] and, therefore, independent of the moving body. Once all the properties of conscious experience are formed in the complex information processes of the brain, the concept of self becomes fixed to a location as a whole experience.[14] What we experience as selfhood is, like this, an internal shadow cast along the walls of Plato's cave. What is *out there* is rendered a representation of the real objects the body encounters but cannot fathom in terms of the far greater abundance of properties these objects possess in actuality.

We might say that the Ego Tunnel's sense of its own self comprises *relations of interiority*; that is, it is a self composed of essential properties mustered together in a location to represent *being* in the world. To know itself, its environment, and the others that populate this environment, the cave brain requires that these properties be fixed in advance. This is a predictive model of inner brain space that requires access to an outside. Indeed, before the Ego Tunnel can attend to its own body image as a whole, it needs to be able to attend to the world—to draw things *into* consciousness.[15] It becomes, like this, a conceptual essence of selfhood, embedded in the brain, so that a self-model can act in the world. However, despite this attentiveness to the world, the Ego Tunnel's solipsistic relation to the outside is firmly limited to experiences *inside* the cave. Certainly, notwithstanding his appeal to mirror neurons as an explanation of how empathy might work—how, that is, the feelings of the other become part of the brain function of the self—this selfhood never really ventures outside of the cave. Following the engineering metaphor, we see that the world outside does not belong, as such, to the software user but to the virtual reality machine itself.[16] The outside is imagined, virtually, as such. This is seeing with *eyes closed,* without windows. It is an outside imagined as a phantom representation, separated from the body, the other, and the environment. In short, this is a virtuality that lacks an affective encounter with the outside. The projections of the world on the walls of the cave are a relation to interiority. Unlike assemblages, they have no *relation to exteriority.*

Another problem with the Ego Tunnel is that it rests on a series of neuroscientific "background assumptions" that are not well supported by the obligatory neural correlates demanded by neuroscientific research.[17] Certainly there is a tendency in Metzinger's theory toward fMRI phrenology in which experiences of the world become embedded in the brain as transparent images of the body as a whole. Despite the assuredness of these locationist inclinations, the idea that the Ego Tunnel is something that is both symbolically and temporally fixed in a space inside the brain amounts to little more than neurospeculation. The brain image has not given up the secrets assumed to be localized concepts and psychological states (imagined or real). The idea of the object (real or imagined) is not traceable to the subject's brain. This embeddedness disregards how the brain–somatic relation may conceptually respond to environmental adaptation and modulation in what Shulman identifies as a delocalized mode.[18]

Instead of looking for concepts inside the brain, there is a need perhaps to explore the sensory worlds in which brains become situated. What follows uses lab rats and mice as aesthetic figures to imagine how sense making is assembled in their world environments through nonlocalized and nonrepresentational means.

THE LAB RAT AWAKENS

A rat wakes up in a dreamy state. He finds himself in a pitch-black cave. Deprived of light, he becomes disoriented, losing his perception of depth. He cannot discriminate shapes. There's no difference between him having his eyes open or closed. The boundary between his body and the space he occupies becomes lost. Although this might still be part of his dream, he feels scared and isolated. Then, suddenly, one of his whiskers brushes up against something hard and cold. He moves his head so that his other whiskers can rhythmically move over this stony surface. One by one, his whiskers become stimulated as he senses what feels like the wall of a cave. He begins to move along it, still using his whiskers to position himself. Soon he senses that he is not alone. His whiskers brush past another rat, then another, and so on. Before long he becomes part of a mischief of rats using their whiskers to edge their way along the wall of the cave, in hope that it may lead to the outside. Eventually, though, the rat wakes up from his dream. He is not in a cave after all. As he stirs into wakefulness, he finds that his head has been strapped inside an fMRI radiofrequency coil.

He has been anesthetized to prevent him from moving in the magnet while data are extracted from his brain. An image is taken of his head in the absence of any stimulation to his whiskers, followed by an image corresponding with one of his whiskers being tweaked, and so on, until all his whiskers have been mapped to a neuroimage.

These kinds of experiments with rat and mouse whiskers provide an intriguing map of the neurocorrelates established between sensations and localized neuronal activations (see chapter frontispiece), the like of which have also been activated in humans, according to visual stimuli relating to, for example, motion, color, and intensity.[19] They provide a compelling indication of the localization of sensation in the brain and have, not surprisingly, encouraged researchers in the neurosciences to assume that cognitive, emotional, and intentional mental concepts may be similarly localized brain activities.[20] That is to say, the brain image of energy consumption and blood flow in the rodent constitutes a mental state. Are we to assume, then, that the rat's whiskers correspond to a phenomenological model (real or virtual), already there, inside his brain? This way, what is outside will become (still following the metaphor) encoded *inside*. But are we not, once again, in danger of ignoring Bergson's point about assuming that the shape of the nail gives us the shape of the coat? To be sure, although fMRI imaging has pointed to localized stimuli being attended to, the process by which what is attended to becomes a concept is highly conjectural. Although neuroimaging has in many ways revived the fortunes of cognitive psychology, and its belief in the existence of internal mental representations of the world, concepts are not proven to be uniquely localized phenomena.

The rodent is assumed to use his whiskers to explore new sensory environments, but the material trace of this stimulation in the brain can be observed as nothing more than an increase in glucose consumption in discrete regions of the brain and linked to behavioral responses.[21] The tickled whisker does not provide an imprint or indeed a see-through representation of the cave located in the brain. At best, as Shulman argues, what is retained is apparently spread over many brain regions that are not reproducible across contexts.[22] Like this, Metzinger's appeal to representational modeling weakens the neural correlates of baseline energies to observable external behavior. In short, he makes an assumption about there being mental stuff living inside the neuron. What fills up the phenomenological space of the rat's brain is not at all self-evident.

Nonrepresentational accounts in the neurosciences are, albeit of a generally behaviorist disposition, compelling. But, if we are to avoid the inside–outside relation altogether, then the rat's intention cannot be considered in terms of strictly cognitive or behaviorist science. Instead, following Deleuze, and Whitehead before him, what is assumed to be on the inside cannot be separated from what is on the outside. The experimenter cannot be separated from the experiment. Both are present in the world, simultaneously. To put it another way, what observes (the subject) is no different from what is observed (the object). This is not a brain that perceives itself from inside the cave but rather a brain that surveys in the folds of the surface.

THE FOLDS OF THE BRAIN

Deleuze argues that folds establish much more complex relations between matter and what emerges from it.[23] To begin with, folds do not separate the matter that is inside from what emerges outside. Descartes's error, which continues to persist in many ways in the neurosciences today, was to imagine that bits of matter can become separated from each other.[24] The waves of the ocean brain do not, as discussed in chapter 1, flow from the inside to the outside. Similarly, sensations do not flow from hardware to become representations in software. Flows do not go

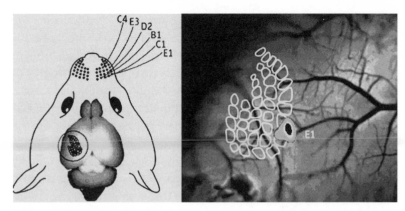

The neural correlate between a mouse's whisker and brain region. Reproduced with permission of Professor Thomas Knöpfel, Imperial College, United Kingdom. Black-and-white version adapted by the author.

from one level to the next; they pass between matter and what emerges from matter. Like this, there is no whole ocean that is not compromised by the outside. Its parts are ceaselessly flowing outside: rainfall flowing down mountain streams to waterfalls, cascading into rivers before entering a sea of parts, which then continue to divide up to form baroque swirls within swirls, currents within currents, and waves within waves.

If, like Metzinger, we were to consider the brain to be a cave, then we would not find it sealed off from the world. Whereas the doors and windows in Metzinger's cave are always shut from the inside, in the fold, they are always opened (and closed) from the outside. We will not have to emerge from one of these doors to tell everyone the bad news that we have all been living an illusion, but rather we would need to impart the good news that the cave is coupled to further caverns, just as the lab rat's brain becomes coupled to the sensations of the cave. Indeed, if we continue to follow the fold, it goes straight through the facade of the sealed cave, passing through the interiority and the exteriority of the walls. It is like this that the fold passes through matter and what emerges from it: the concept of the self, the soul.[25] As the rat discovered, the cave was not an empty space. Likewise, the brain is not an empty space. There is no such thing as an empty space. Regardless of whether the brain is an ocean or a cave, or a mixture of the two, each part of that ocean, or cave-world, contains another world. The rat has the sensation of a cave in his head. This is not the same as saying there is ascension from the sea of parts to the ocean as a whole, a sum total of who we are. The sensations of the cave occupy the rat's brain. Equally, the empty cave is open to adjoining caves that link it to other worlds, including those of the ocean. No cave is sealed off from the threat of flood. The fluidity of matter will eventually find a way through the porosity of the cave walls.

One of the effects of sensory deprivation is the out-of-body experience (OBE)—in effect, to lose the boundary between the body and the outside space. Metzinger's experimentations with OBEs led him to believe that the emergence of selfhood begins in the brain, independently of the body, bringing about the virtual sense of ownership and control. But there is no flight of steps rising from the lower levels of the cave to some higher level from which we can look down at ourselves. The fold has no staircase leading from the many parts to the "eminence of the One."[26] What emerges is neither brought together as One nor lost in the chaos of cavernousness of parts. Even if the parts make up a territory (a sense of self), they remain heterogeneous. They are a multiplicity. The multiple is, like this, "not merely that which has many parts, but that which is folded in many ways."[27]

Finally, Metzinger provides a concept of the hypothetical mirror neuron function that endeavors to open up the cave brain to the social world; that is, an empathic Ego Tunnel emerges when certain layers of the self-model function as a *bridge* to the social domain.[28] But this appeal to the world of exteriority is a phenomenological trap. It is a bridge that is always followed and preceded by a tunnel leading back to the "internal image of itself as a whole."[29] By forcing the mirror neuron hypothesis into the dream tunnel, Metzinger fails to grasp the potential of empathy as an outside force, a relational encounter that Tarde captured so well in his notion of imitative cell-to-cell communication—how, that is, the rat's encounter with the sensory environment of the cave, including the associations he makes to other rats, draws him into the multiplicity of the imitative crowd. This is the rat becoming a crowd brain. As follows, the Ego Tunnel misses the crucial role of relationality in multiple processes of subjectification. This is not just a philosophical problem; it is, evidently, political. The discussion here therefore needs to shift radically from the inner model of self to how we can understand subjectivity in the making steered by the brain–somatic relation to the spatiotemporal demands of the sensory environment. This is an appeal to the fold instead of the tunnel, which further opens up political questions concerning intentionality, not in terms of storage or location, but understood as ensnared in imitative processes of subjectification and subject to an increasingly managed intensification of sensory stimulation and deprivation.

RADICAL RELATIONALITY AND RHYTHMIC IMITATION

The politics of the fold can be grasped by bringing together two approaches (Protevi's radical relationality and Borch's rhythmanalysis), which (1) work to erode the border between an inner self and an outer sensory environment and (2) place the temporal rhythms of Tarde's imitation repetition as the base of social relationality. To begin with, and following Protevi's reading of Bruce E. Wexler's book *Brain and Culture*, we see how the inside–outside relation is substituted for "radical relationality"; that is to say, "being human is composed of relations; we do not 'have' relations, but we are relations all the way down."[30] In what follows, radical relationality helps to theorize the force of the outside in a number of significant ways:

1. Wexler notes how neuroplasticity becomes open to varying degrees of change, over time, occurring in a *neuroenvironmental emergentism*; that is, the intricate connections and patterns established between neurons are "determined by sensory stimulation and other aspects of environmentally induced neural activity."[31]
2. Radical relationality reverses Metzinger's process of virtuality, making the essence of the self, not the objective reality the self encounters, illusory. In other words, the sense of self (a substantive part of who we are) is the "imagined" outcome of the speed of sensory processing being too slow to perceive anything more than the self as an individuated substance embedded in the brain. So, rather than rendering the brain an individuated substance, bequeathed with fixed properties, Protevi contends that subjectivity is made in the "tendency to partake in a pattern of social interaction." It is not, therefore, the virtuality of the individuated self that determines how the social field is perceived; rather, it is "the interaction of intensive individuation processes that forms the contours of the virtual field."

3. Wexler's radical relationality forces this substance viewpoint (from the inside out) to a new perspective in which what is internalized is a "pattern of interaction." This is radical because, as Wexler writes,

> the relationship between the individual and the environment is so extensive that it almost overstates the distinction between the two to speak of a relation at all.[32]

Bodies and brains are in constant processual exchange with the environment, which, although appearing to be masked by an individual's "exaggerated sense of independence," carried in a fleeting memory that considers our uniqueness to be a property of who we are, nevertheless makes us little more than an effect of the sensory environment.[33] Therefore, what Protevi significantly extracts from Wexler's plastic brain thesis is an emergent subjectivity understood not as the outcome of complex malleable brain functions but from a "differentiated system in which brain, body, and world are linked in interactive loops." Wexler's entire project is consequently underlined by his intention to

> minimize the boundary between the brain and its sensory environment, and establish a view of human beings as inextricably linked to their worlds by nearly incessant multimodal processing of sensory information.[34]

4. In addition to Protevi's reading, Wexler foregrounds the ubiquity and automaticity of imitative processes as key to understanding subjectivity in the making as an effect of the sensory environment.[35] As Wexler puts it, imitation is "consistently operative throughout the moment-to-moment unfolding of everyday life."[36] It is through the close bonds a child makes with a range of caregivers that the imitation of example persists through social relations. The extent to which the imitation of example occupies the interactive loops that compose subjectivity in the making suggests a distinctive Tardean aspect to Wexler's sensory environment.

One way in which to expand on the imitative quality of radical relationality is to follow Christian Borch's observation of the rhythmic intensity of Tarde's imitation suggestibility, that is to say, to take note of "how rhythms simultaneously condition imitations and produce ruptures," further noting how these rhythmic relations in the social environment pass through the three parts of a Tardean diagram (imitation repetition, opposition, and adaptation).[37] A number of aspects of Borch's rhythmanalysis coincide with radical relationality and can therefore be taken forward in the context of this discussion. That is, first, what comes together through imitation repetition is not the unity of the One; rather, rhythm produces the harmony of the many—a harmonious relation between repetition and difference. Rhythm does not therefore produce the stability of a self-contained spatial identity—a permanent self fixed in a sum total, or systematic emergence of a whole from which the materiality of the brain emerges on the outside (a soul, a mind, a person, a model)—but rather, second, denotes materiality in rhythmic movement. Again, this is not a flight of steps that leads from the microlevel of the cave to the macrolevel of consciousness, or from the many to the One, but a folding rhythm of movement through a sensory environment of oppositional collisions and social adaptation. In this way, Borch further demonstrates the significant role of rhythmic imitation in the political positioning of protosubjectivity, that is to say, the situating of subjects in spatiotemporal flows of the sensory environments they inhabit. As Borch puts it,

> the individual does not exist prior to the rhythms but, on the contrary, is produced by them and their momentarily stabilized junctions, and since the subjectification of the individual therefore changes as the rhythms and their junctions change, rhythmanalysis is not merely a perspective on imitations per se, but equally a tool to demonstrate a society's dominant ways of promoting subject positions.[38]

Indeed, Borch draws attention to Tarde's observation that environments, like rural communities and newly industrialized cities, acquire a "very significant importance in what is actually . . . imitated."[39] In Tarde's era, cities produced new social formations, like the urban crowd, which, unlike rural family communities, become powerful vectors for imitative flows of the inventions of fashion, crime, and, potentially, nonconformity and riotous assembly, for example.

In times of rampant capitalist industrialization, the revolutionary capacity of Tarde's crowd brain is transformed by the intensification of mediated relations in urban environments. This social adaptation begins with the rise of the press, which ensures that the crowd becomes increasingly "disconnected from [the] physical co-presence" of the urban environment.[40] Here we encounter Tarde's nineteenth-century media theory in action. He would indeed upgrade the crowd brain's rhythmic encounter with the city to incorporate the introduction of mediated communications that create new publics, anticipating, in many ways, the progressive onset of the sensory environments of the mass media age. Newspapers, cinema, radio, and television become component parts in the emergence of these Tardean publics. To be sure, the trajectory of these newly mediated populations extends to contemporary post-industrial sensory environments that play a significant role in once again rupturing harmony and repositioning subjects in the rhythmic flows of digital culture—although seemingly deterritorializing the copresence of the crowd brain, the quickening rhythm of digital culture brings together (or reterritorializes) cells, brains, people, crowds, publics, masses, and ubiquitous technologies into nascent sensory terrains.

Significantly, this focus on newly mediated sensory environments returns us to Stiegler's concerns regarding the problem of how to care for a generation whose attention is increasingly steered away from intergenerational caregivers toward sensory experiences infiltrated by neurologically enhanced systems of marketing power, that is to say, environments controlled by the carelessness of capitalist neuropower. Indeed, my intention in the concluding part of this chapter is, at first, to acknowledge the significant role of caregivers in managing the rhythmic flows of imitation and, last, to consider alternative care systems with the capacity to respond to a progressively more rhythmically controlled sensory environment.

THE CAREGIVER

It is Wexler who draws attention to the role of significant caregivers in the operation of imitative encounters between a child and the sensory environments they pass through.[41] A number of layers in this relation need to be considered. First, the infant's primary sensory experiences are the caregivers he or she encounters. It is these caregivers who direct the rhythm of imitation, steering attention, feelings, and thoughts about it.[42] Second, as Protevi notes, the relation between caregivers and children provides a "scaffold" for learning what is just "beyond" what the child is "capable of at any one moment," as a consequence allowing the infant to internalize this support.[43] Like this, it is the regulatory powers of the caregiver that shift toward the developing self-regulation of the infant in the sensory environment.[44] Third, it is important to acknowledge the infant's protosubjectivity, namely, a crude capacity to imitate. It is indeed the internalization of this scaffold that interacts with the external environment. But, as Wexler's radical relationality maintains, the interactive looping of what gets imitated intensifies to a point where any boundary between the brain and the sensory environment becomes indistinguishable.

Given that the caregiver's role of providing a scaffold to support the infant's future experiences of sensory environments is supposed to diminish over time, attention is drawn toward what kind of care systems replace the primary caregiver. It is not enough, however, to assume that when a child grows into an adult, she becomes a scaffold herself. Adolescents and adults continue to be exposed to the increasingly marketized forces of the sensory environments they inhabit. Indeed, the transition from adolescence to adulthood is arguably being extended through the many appeals made to infantile desires for games and other entertainments formerly targeted at children. A question needs to be asked concerning the nature of the new imitative rhythms that position subjects. To answer this question, I will offer two points of intervention. First, following Stiegler to some extent, surrogate caregivers, that is, schoolteachers, need to be seen alongside the antagonistic influence of systems of carelessness (marketers, the media, and workplace managers, for example), which endeavor to tap into the looping imitative interactions established between managed sensory environments and the "irreversible" structuring of a child's "synaptogenetic circuits."[45] Second, then, there is a need to expand on the notion of the caregiver to consider a broader concept of *systems of care,* namely, systems that intervene in the rhythmic flows between brains and sensory environments, providing alternative scaffolds. These include, on one hand, the *bad* sensations of a biopolitical apparatus, which, as Stiegler contends, captures youth attention initially to destroy it before reinventing it in the service of the market, and, on the other, alternative care systems that might offset, or even neutralize, in a positive way, the frequency-following disciplines of capitalism. Clearly this opposition between good and bad sensory environments is complicated, but for the moment, let's stay with it.

GOOD SENSATION, BAD SENSATION

A further question needs to be asked concerning what happens when the caregiver's scaffold acquiesces to mediated sensory environments that coincide with neurocapitalism. In other words, what kind of care-lessness materializes when human synapses and the digital network fuse together with neuromarketing power? This is a problem that can be approached using a combination of Wexler's radical relationality and Borch's Tardean rhythmanalysis. As we have already seen, Wexler advantageously helps us to conceive of the imitative relations established between youth and sensory environments as being without boundaries, therefore grasping neuropower as an outside force. However, although Wexler's approach also provides a plausible explanation of how systems of care, and carelessness, persist by way of, it would seem, sensory stimulation and deprivation that encourage anxieties and addiction, ultimately, his thesis lacks an engagement with the political implications of these two opposing poles. It is here, then, that Borch's account of a Tardean rhythmic imitative encounter with the environment offers a much-needed political dimension to this discussion.

To begin with, Wexler's thesis distinguishes between sensory stimula-tion and deprivation. On one hand, environmental stimulation allows the brain to develop the capacity to act on the world,[46] whereas, on the other, withdrawal of sensory input impairs brain development[47]— not surprisingly, perhaps, because brain development is, according to Wexler, determined by the quality and quantity of sensory stimulation. Solitary confinement in a dark cave is a cruel punishment, for rat and human brains alike. What would become of the rat in Plato's cave without his whiskers? The growth of cell-to-cell connectivity depends, it seems, on the interactions between active brain cells and sensory environments (e.g., when a rat has some of its whiskers clipped, neuronal resources move to the remaining whiskers). Even when the clipped whiskers are allowed to grow back, these resources never return.[48] Such deprivation would lead the rat to seek out stimulation by any other means. Like humans deprived of stimulation, he would perhaps become depressed, anxious, even suicidal. Significantly, there is a physical craving for sensory stimulation. The human, like the rat, would seek out a more advantageous place that might provide some relief for her addiction to good sensations.

If the desire for stimulation were to remain unsatisfied in the darkness of the cave, the rat and the human would begin to hallucinate, not necessarily phantom limbs, or whiskers for that matter, but light, spots, shapes, objects, forms, other humans or rats.[49] These hallucinations need not necessarily be visual. Auditory sensations are experienced when human subjects are deprived of sound and light. These hallucinations are not dreams either. EEG tests carried out on human subjects kept in the dark and deprived of sound for seventy-two hours indicate that visual and auditory hallucinations occur when people are wide awake.[50] Here we find another clear distinction between Metzinger's cave brain, which hallucinates itself, and Wexler's radical relationality, in which it is the sensory environment that is hallucinated. This is a force of the outside that ensures that the cave is never sealed. To be sure, hallucinations in which other humans and animals appear suggest that there is a strong desire to reach out to the social world. Sensory deprivation increases the desire to reach out to others for stimulation. In other words, the socially deprived brain craves interaction outside of itself. This is the rat's reward for becoming social. Similarly, the desire for stimulation becomes central to the looping imitative interactions between the child and the caregiver. Physical body-to-body interactions, like gestures and facial expressions, are imitated and passed on through intergenerational cycles. It is, like this, that the caregiver furnishes the world in which stimulation is experienced, and acted on, with attention-grabbing moments of fascination and curiosity.

Importantly, the imitative structuring process needs to be seen as something that occurs beyond the inside–outside relation between the child's synaptogenetic development and the externally managed world. The intergenerational conditions of attention formation, as Stiegler refers to it, become a folded loop of radical relationality spanning generations.[51] Like this, interactions between external sensory stimulation and internal scaffolds need to pass through generational cycles of imitation, wherein the child is opened up to external forces (brain-becoming-adolescent) testing out internal structures on the external world, before, in turn, the scaffold of the adult becomes the new sensory fascination of children, albeit, as Wexler contends, becoming more fixed, conservative, and antagonistic to difference with age.[52] This is indeed a folded intergenerational relation that Stiegler extends to the dead. The formation of attention to the world requires, as such, that living ancestors become "transmitters of experience accumulated across many generations, connecting the child with *dead* ancestors."[53]

THE NEUROPOLITICS OF SENSORY ENVIRONMENTS

Despite providing a valuable rethinking of the brain's radical relation to the sensory environment, Wexler's caregiver needs to be disentangled from the discursive formations that surround neuroplasticity. This is because these formations are all too often engaged in a crude neuroscientific targeting of the brain that obfuscates the relational capacity of sense making to be free. There are three examples that can help to explain how these discursive forces work in sensory environments. First, we need to address the processes of normalization that care systems introduce, particularly those in which neuroplasticity, as presented in neuroimaging, supports conservative policies that determine what constitutes good or bad caring. Second, the binary tendency in Wexler's focus on good stimulation–bad deprivation obscures the complex affective politics apparent in radical relationality. The point is that the dynamics of joyful affective encounters need to be understood, as Protevi contends, in their registering of passive and active power relations. The joyful encounter is a political tool that has historically rendered large percentages of the population docile in the face of fascism. Last, and relatedly, I want to intervene in Wexler's notion of an apparent inhibition of intergenerational desires for novelty. That is to say that brains seem to have a preference for familiarity over difference. Indeed, there is a marked tendency in neurocapitalism for brains and bodies to become entrained by the rhythms of the sensory environments they encounter, and, as a result, they may become more inclined toward the kind of conformism Gramsci warned about. This discussion therefore concludes by beginning to think through alternative rhythms of imitation that admit and encourage a progressive diversity of sense makers outside the neurotypical.

Sensory Deprivation in the Age of Austerity

A critical theory of sense making needs to address the relation between neuroscientific discourses concerning brain plasticity, sensory deprivation and stimulation, and government policies that seek to determine what constitutes normative care. Here we again encounter the discursive influence of the locationist tendencies in neuroimaging. In this respect, the journalist Zoe Williams draws important attention to the controversy surrounding the so-called First Three Years movement.[54] This is a political movement that purposefully misinterprets and distorts neuroscience research to (1) assign blame to individual parents, who are often living below the poverty line, and (2) justify state intervention into parenting intended to produce a certain kind of subject. As the neoconservative politician authors of *Early Intervention: Good Parents, Great Kids, Better Citizens* contend, "neuroscience can now explain why early conditions are so crucial."[55] That is to say, positive stimuli concretely correlate with the development of brain cells and synapses in a "finite time window for learning certain things."[56] These are state-determined care systems very much shaped by, Williams argues, the persuasive wow factor of the neuroimage:

> We're not actually seeing inside brains. We're certainly not seeing emotions written on to the brain that we can then draw conclusions from into how parents should love their children.[57]

Moreover, though, the neuroimage used to support the report is taken from research into extreme neglect by way of sensory deprivation (i.e., removal of light and human contact over a long period of time). It has no relation whatsoever to the experience of being brought up in a *broken home,* nor is it linked to a tendency to commit crime later on in life, for example. However, taken out of context, what these neuroimages do is determine a discursively shaped logic of what normalized care should look like:

That first bond between child and care-giver determines everything that comes after; if it's disrupted or inadequate, nothing can
really repair that.[58]

The use of locationist neuroscience becomes part of the neoconservative discursive formation when the report draws attention to the effects
neglect can have on particular regions of the brain. For example, the
limbic system, which the authors note "governs" emotions, can, it is
claimed, shrink by up to 30 percent if neglected. Similarly, the hippocampus ("responsible for memory") can also wither owing to "decreased
cell growth, synaptic and dendrite density—all of which are the direct
result of much less stimulation."[59] Indeed, this is neuroplasticity grasped
through the lenses of the cybernetic metaphor of cognitive neuroscience.
As the report states,

just as a memory will fade if it is hardly ever accessed, unused synapses wither away in what is called "pruning." In computer terms
what takes place is the software (programming by the caregiver)
becomes the hardware (the child's fully-grown brain). The whole
process has the effect of making early learned behaviour resistant
to change.[60]

The notion that babies who fail to make the right neural connections in
a finite three-year period become lost to a system of care unsurprisingly
feeds into a familiar conservative discourse concerning intergenerational
cycles of deprivation in which certain individuals (predominantly un-
employed or lower-income groups) do badly at school, lack empathy,
have mental health problems, and, naturally, succumb to criminality.
This is a discursive formation that maintains that internal synaptogenetic
developments determine social mobility *in isolation* and that government policy needs to make these developments central to family law
and child protection policies, which can filter out, as the report puts it,
"the supply of dysfunctional people to manageable levels."[61]

What is clearly amiss in this discourse is the caregiver's lack of access to resources, like a provision of adequate nutrition, which wards
off brain-damaging malnutrition, or sufficient financial support to
pay for good quality early-years, and lifelong, education. In short, the
early-years intervention is a care system made in the conservative age

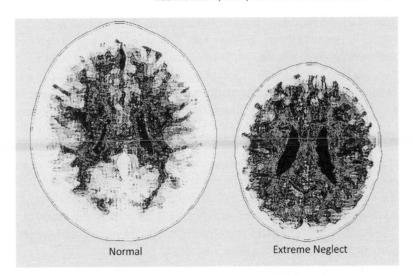

Normal   Extreme Neglect

The author's impression of the neuroimage used in the *Early Intervention: Good Parents, Great Kids, Better Citizens* report published in 2008. From B. D. Perry, "Childhood Experience and the Expression of Genetic Potential: What Childhood Neglect Tells Us about Nature and Nurture," *Brain and Mind* 3 (2002): 79–100. Permission to use the original image was refused owing to the controversy caused by its appearance in the report.

of austerity. It situates the problem of caregiving solely in the lap of parental responsibility, without providing appropriate support for the costs of care. Ultimately, it is not the brain but the social relation to the world, experienced through impoverished affective atmospheres, that is sick.

The Passivity of Joyful Encounters (1923–38)

Wexler's positioning of stimulation as *good* and deprivation as *bad* is significantly complicated by Protevi's reading of *Brain and Culture*, which draws attention to the political dimensions of affective encounters with sensory environments. To fully grasp the importance of this intervention, we need to begin by briefly following Protevi's use of the Spinozan constituents within Deleuzian ontology to build on the concept of radical relationality—how, that is, the affective states involved in the looping interactions between child and caregiver can be reconceived of as relational desubjectified processes. The emotional contagions that

Wexler establishes, for example, between the child and caregiver are regarded not as the transference of an embedded substance from one person to the next but rather as a relational "interindividual process."[62] A panic contagion, for instance, has an affective capacity that can exceed the persons affected by it. To be sure, crowds have the capacity to panic collectively. So, although the affect of a panic is "felt" by the individual members of the crowd, the affective contagion itself is not the same as a subjective feeling. The feeling of affect extends beyond the subjective, so that in "extreme cases of rage and panic," it triggers "an evacuation of the subject as automatic responses take over."[63] Similarly, the transfer of the caregiver's neural scaffold to a child's brain does not equate to an exchange of felt substances (one mapped to the other) but is rather a radical affective relation, which is, in itself, an individuation of a "distributed and differential social field."[64]

Significantly, Protevi's focus on the affective components of radical relationality returns us to the two central questions raised in this book (*what can be done to a brain?* and *what can a brain do?*). Indeed, unlike Metzinger's cave brain, which navigates the world by representing (virtually) features of that world to itself, Protevi offers an affective brain–somatic relation that negotiates the world by way of "feeling what they can and cannot do in a particular situation."[65]

This is the political power of affect grasped in the double event of Deleuzian ontology: the capacity for a body to affect and be affected. On one hand, the passive register (to be affected) is grasped politically as an example of *pouvoir*. It is a bodily encounter in the world, similar in many ways to a Tardean imitative encounter between the somnambulist and power dynamic of an action-at-a-distance. This is a crowd brain desiring its own repression and craving the control of a transcendent fascistic leader. Importantly, this is a desire satisfied by way of joyful as well as fearsome encounters. On the other hand, the active register (to affect) is an encounter determined by what Protevi calls "mutually empowering connections." Political power as *puissance* is equal to "immanent self-organization," "direct democracy," and "people working together to generate the structures of their social life."[66] Is it, in other words, an "active joyous affect," increasing the puissance of the bodies that pass through the sensory environment, enabling them to "form new and mutually empowering encounters outside the original encounter"?[67]

The passive joyful encounter perhaps has its modern political origins in the fascisms of the late 1920s and 1930s, in particular in the Nazi propaganda machine, which thoroughly grasped the purchase of appeals to pleasure as well as fear. The encounters they produced were carefully assembled experiences that tapped into the desires of the crowd. Both Hitler and Mussolini were apparently well acquainted with the late-nineteenth-century crowd theories of Gustave Le Bon, and not surprisingly, they endeavored to draw on his notion of hypnotic mass suggestion as a mode of control. To be sure, the many direct appeals to desire fit squarely with Tarde's more exacting microsociology of the crowd, particularly his idea that the object of the desires of the social sleepwalker is always belief.[68] That is to say, to make a population believe in fascism, it was necessary to appeal directly to desires for joyful sensations and to maintain atmospheres of absolute terror. Like this, the large-scale state-run leisure organization Kraft durch Freude (Strength through Joy) demonstrated how the Nazis placed a heavy emphasis on the happiness of the population and its desire have a good life so that they would associate these feelings of joy with the new order.[69] This was not so much an ideological trick working directly on belief systems as it was an attempt at tapping into the crowd's vulnerability to mass suggestion experienced through joyful encounters.

Thinking through the oppositional tensions of this double event, Protevi raises important questions concerning the multiple processes of subjectification occurring in politically organized affective encounters with the Nazis at the Nuremberg rallies. These large-scale militarized events provided a stimulating sensory environment that can be ethically gauged according to a kind of pouvoir that elicits passive joy, while, at the same time, enforcing the rhythmic entrainment and repression of the crowd. As Protevi puts it,

> the Nuremberg rallies were filled with joyous affect, but this joy of being swept up into an emergent body politic was passive. The Nazis were stratified; their joy was triggered by the presence of a transcendent figure manipulating symbols—flags and faces—and by the imposition of a rhythm or a forced entrainment—marches and salutes and songs. Upon leaving the rally, they had no autonomous power (puissance) to make mutually empowering

connections. In fact, they could only feel sad at being isolated, removed from the thrilling presence of the leader.[70]

The marketers of early fascist joy also understood that conventional party politics, or indeed totalitarianism, was never going to be something that the population desired. It is much better to appeal to the desire to oppose the established political order than it is to appear to personify it. In the early years of Italian fascism, Mussolini purposefully positioned his fascism as the "antiparty" so as to appeal directly to the disaffected classes' desire to disrupt.[71] This is an example of a passive joyful encounter because it seems to offer power to those without access to political resources. However, despite initially appealing to productive desires for change, fascism of this kind does nothing more than exacerbate the repression of the masses.

Right-Wing Populism: Waking the Somnambulist in 2015–16

Farage . . . comes along and people connect to him because he sounds like the guy in the street.
— Canvey Island Independent Party member explaining the appeal of the U.K. Independence Party's leader in Essex[72]

The disempowering encounters with Nazi joy are comparable in many ways to a fascistic trajectory persisting in current waves of right-wing populist contagion spreading throughout Europe and the United States at this point in time—a disparate series of political movements that similarly position themselves as antiparties opposed to the established order. Once again, these attempts to position far-right politics as a radical movement add up to more than a mere ideological trick by a totalitarian military machine. There is a far more complex and subtle relation established between desire and belief: a relation that has many continuities and discontinuities with the past. To begin with, although the entraining rhythms of marching and salutes have, for the most part, faded into the background (for the time being, that is), the entrainment of the population by way of affective appeals to feelings about nationhood, race, and unity persists. Moreover, this is a right-wing populism stimulated by affective encounters intended not only to destroy difference and celebrate sameness but also to produce repression through joy. Not surprisingly, then, Wilhelm Reich's question concerning why it is that so many people seek their own repression under regimes with political motivations that are palpably counter to their own self-interest has been revisited. As Reich put it in 1946,

> what was it in the masses that caused them to follow a party the aims of which were, objectively and subjectively, strictly at variance with their own interests?[73]

Indeed, we need to ask why, again, today, so many people desire pouvoir over puissance. They seem to be wide awake. They do not appear to be deceived. But it is not freedom that the sleepwalking supporters of right-wing populism desire; it is repression. They are, once more, in need, it would seem, of a transcendent authority to protect them from what they are told is the chaos of an economic depression worsened by porous national borders open to floods of virus-ridden immigrants stealing jobs, scrounging welfare, and intent on acts of terror. That is, as well as having someone to blame for their own disempowerment, they crave an authority figure to relate to, someone who personifies prejudicial beliefs and anxieties stirred into action by a fear of the unfamiliar. So where amid all these appeals to the fear of otherness is the joyful encounter? To answer this question, there is a need, on one hand, to rethink the sex-economic sociological framework in which Reich framed his original question, that is to say, to move on from its recourse to the inner world of an unconscious mind rooted in biological drives, and address, instead, the affective relations established between the population and the sensory environments that situated it. What seemed to Reich to be the perverse impulses of the fascist unconscious—a desire for repression of biological impulses that seeps through the layering of the unconscious into conscious rational choices—needs to be revisited in terms of a political affect that stirs into action a different kind of mass somnambulism. This is not a hidden unconscious seeping out from the inside. Affect is not a fantasy from within. The sleepwalker is already *out there*, in the crowd—the guy in the street. The somnambulist is a social relation. This is the kind of microfascism that is not simply personified by a transcendental leader, either; as Michel Foucault notes, it is already in "everyday behavior"—it is "the fascism that causes us to love power, to desire the very thing that dominates and exploits us."[74] On the other hand, there is perhaps a need to revisit certain elements of the critique of Marxism Reich offered in the 1940s. Contrary to how the masses have been generally observed through the lenses of Marxist theory, the working-class supporters of these right-wing movements did not appear, as Reich argued, to perceive themselves as a hard-done-by proletariat in opposition to bourgeois elites. As Reich contends, the working classes of the 1940s did not see themselves as the struggling class anymore. They had, he claims, grasped themselves as having "taken over the forms of living and the attitudes of the middle class."[75]

Today, it would seem that the supporters of right-wing populism have become particularly susceptible to differently oriented appeals to the felt experiences of a shifting sense of class identity than those Reich observed. That is, the working classes are now positioned as the disaffected guy in the street. This means they are once again drawn to the appeal of the antiparty, because it seems to soak up a desire to disrupt order but merely produces more repression. Moreover, desires are now shared in the digitally mediated sensory environments of social media, these digital crowds and data assemblages Facebook readily experiments with. Indeed, Obsolete Capitalism's analysis of the rise of comedian Beppe Grillo's popularist antiparty in Italy, the Five Star Movement, points to the emergence of a digital populism that acknowledges the central role the marketer and net strategist have in building the antiparty's brand, orienting voters, and disrupting dissidents through social media.[76]

In many parts of Europe, there is a distinct reversal of the fortunes of Reich's imagined upwardly mobile proletariat, which the right-wing popularists in the United Kingdom are readily exploiting by way of joyful encounters. This is again not simply a trick of ideology played out on the ignorant masses. Like Grillo in Italy—the authoritarian hiding behind the rascally face of a showman—we find that the bourgeois elites, secreted away behind the facade of these antiparties, are endeavoring to pass themselves off as the guy in the street, or at least some jovial personality compatible with the contrivance of this imagined worldview. The U.K. Independence Party (UKIP), for example, is led by a privately educated former stockbroker who is regularly filmed and photographed by the media sharing a pint of beer in the local pub, creating an appealing impression that he is *one of us*.[77] The production of these political ersatz experiences of joy cannot simply be attributed to an ideological appeal to a rigid sense of the representation of class. The image of Nigel Farage swigging a pint in the local pub works in the insensible degrees between representational illusion and affective states that trigger the desire for mass repression; that is to say, they exist in the *interferences* between the desires and beliefs of a population. To be sure, it is the triggering of these latter affective states that seems to prompt contagious overspills of affect that are as much about joy as they are about fear. Although a large percentage of the affective contagion of UKIP can evidently be put down to racist fearmongering over immigration, it is also the case that supporters of the right become vulnerable to joyful encounters with these showman-like leaders, which have been historically satisfied (e.g., in the United Kingdom) through right-wing inventions like the royal family and Saatchi and Saatchi's fabrication of the handbag-swinging shopkeeper Margaret Thatcher. The fascist marketer has, like this, continued to perpetuate a sensory environment full of joyful encounters with congenial aesthetic figures: right-wing buffoons, including the UKIP leader and the former mayor of London, Boris Johnson, whose jester-like performances obscure the inequity of power relations in the United Kingdom wherein the many are overwhelmingly dominated by an overprivileged and privately educated few. Surely the point is that the somnambulist needs to wake up! These buffoons are not *One* of us. Following Tarde's microsociology, these global leaders should not be grasped as personifications *collectively born*.[78] They are a monadological accumulation of *facts* (e.g., Farage likes a pint and Boris always says it how it is) that tend to assemble and resonate, not so much by accident, as Tarde contended, but by way of purposefully steered affects that spread through the sensory environments of the mass media and digital populism.

Returning to the earlier focus on neurology, the affective marketing of right-wing buffoons can perhaps be seen alongside a more generalized marketing of ersatz experiences in the affective politics of neurocapitalism that stimulate a craving for sensory stimulation. Again, this is a regime of control that asks questions of conventional Marxist ideology. As Reich points out, the ideal of abolishing private property, for example, seems to clash with a mass desire for commodities of all kinds.[79] In the 1940s, Reich listed such mundane items as shirts, pants, typewriters, toilet paper, books, and so on, but today we can add a far more sensorial list of luxury consumer items, including the much-ridiculed

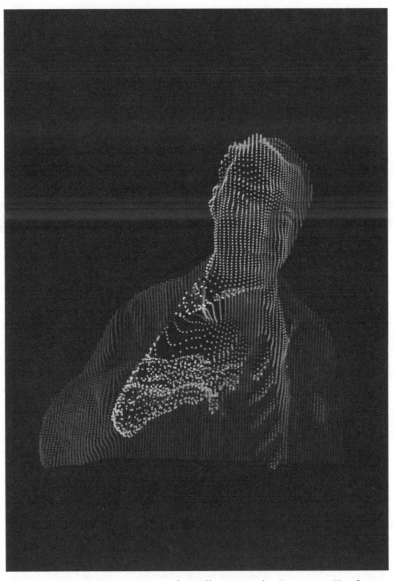

UKIP leader Nigel Farage: one of us? Illustration by Francesco Tacchini.

working-class obsession with wide-screen TVs and access to social media entertainment systems that connect populations to an array of further joyful encounters. These are more than ideological weapons of mass distraction. Indeed, following Bruce E. Wexler's thesis, the desire for joyful sensory stimulations of this kind can be conceived of as an addiction that exceeds the commodity fetish, the satisfying of which reduces the anxieties and depressions caused by sensory deprivation.[80]

In many ways, manipulative incursions into the joyful encounter correspond with a marketing power that disrupts intergenerational cycles of care, ensuring that adults are suspended in perpetual juvenile states of joy seeking and, it must be added, engaged in a perpetuation of addictions to childlike sensory environments provided by the Disneyfication (and gamification) of everything. Indeed, turning Wexler's thesis on its head, it might even be argued that given the anticipated overwhelming control of sensory stimulation by neurocapitalism, some level of deprivation might actually be a good thing.

THE RHYTHMIC BRAIN: FAMILIARITY AND DIFFERENCE

By way of attempting to explain how the kinds of prejudices Donald Trump spreads through a population, Wexler identifies a neurological antagonism to difference that provokes conflict based on dissimilarities in belief systems. This results, he contends, in a continuous battle for control of the sensory environment.[81] In short, people come into conflict with each other because there is a mismatch between external structures and internal scaffolds. They seek, like this, to resist "others from filling their environments with structures and stimulation that conflict with their own internal structures."[82] Following Robert Zajonc's noncognitive theory of affective states, Wexler points to a tendency toward conservatism found in a brain that perceives familiarity as more positive than difference. There is, accordingly, an unconscious preference for the repetition of the same, because familiarity, as Zajonc argues, produces consonance, while difference produces dissonance.[83] Again, in light of Trump's rise in the polls, perhaps some level of dissonance is a good thing. Certainly the role of the caregiver should be to open up the sensory experiences of the child to novelty and difference that disrupt the regime of the same.

Perhaps the rhythmanalysis of the harmonious and discordant relations established between repetition and difference also helps us to approach that other normalizing regime of neurocapitalism: the control of attention. This is because ADHD can similarly be grasped in a tension between the constancy of normative rhythmic consonance and the variability of arhythmic dissonance as established in brain frequency following. Care systems used to treat ADHD are engaged in the rhythmic correction of *anomalous* brain frequencies via neurochemical intervention and EEG. ADHD has become a component of a prevailing neuroscientific discursive formation wherein the problem of inattention and hyperactivity is located *inside* the brain. However, ADHD also highlights the brain's rhythmical relation to the sensory environments in which the inattentive subject becomes situated. This relation is consistent with Borch's theory of the rhythms of urban environments, which set new norms in terms of early industrialized entrainments. ADHD is, like this, attributed to anomalous rhythms that do not fit the quickened tempos of postindustrial sensory environments.

The extent to which the dissonance of inattention and hyperactivity is leading to a failure to synchronize brain waves to these rhythmic demands is perhaps evidenced in a sharp rise in the prescription and use of psychostimulants like Ritalin and Adderall, intended to get people back up to attentive speed while also making them more passive to disciplinary mechanisms. Similarly, EEG technology is fast becoming central to the management of individuals who find themselves in increasingly demanding and controlled sensory environments. To begin with, EEG is used to measure the effects of psychostimulants on schoolchildren, wherein alpha and beta frequencies paradoxically increase in chorus with each other, leading to heightened attention and docility. An EEG test for ADHD was also approved for marketing in the United States last year.[84] The test correlates brain wave frequencies with a child's capacity to pay attention in the classroom. In short, inattentive subjects are measured according to a higher ratio between theta and beta frequencies associated with depression, hyperactivity, impulsivity, inattentiveness, daydreaming, and poor cognition. Moreover, used as an alternative to medication, EEG-driven brain wave stimulation is used to synchronize neuronal activation and intensify the efficiency of the synapses to respond to increasing demands from the sensory environment. Changes made to brain wave frequencies equate to changes in behavior, as rated by teachers and parents. Significantly, what amounts to the regulation of temporal neural activity does not, it seems, take into account the need for rhythmic brains to adapt to the increasing sensory demands of neurocapitalism. There is, for example, no recognition of the neurodiversity of brain frequencies in the development of either chemical- or EEG-based care systems. These tests follow instead the medical model of disability, which implies a "neurotypical society" beset by anomalies, dysfunction, disorder, symptoms, and comorbidity.[85]

Following Borch, but also E. P. Thompson's temporal study of industrial capitalism and Paul Virilio's notion of speed factories,[86] these attempts to normalize the rhythmic brain can be further understood as endemic to a neurocapitalism engaged in a frequency-following discipline, that is to say, a mode of capitalism that demands that a rapid cadence of attention and passivity be controlled by neurochemical and technological invention. These mechanisms of control further coincide, to some extent, with Stiegler's concerns that attention deficit is an upshot of a battle for youth attention between marketers and intergenerational systems of care. As Stiegler argues, youths today are being turned into *attention engines*: a product of the carelessness of a rampant industrial populism that requires individual attention to be "particularizable," "formalizable," "calculable," and "controllable."[87] However, it is important to grasp how radical relationality reconceives of the mediated encounter between caregivers and the child in these sensory environments in ways that avoid the impasse of the inside–outside relation. In short, it is important to acknowledge how the infant's assemblage brain, expressed through a capacity to imitate, is established in the interactive looping of what gets imitated. It is the rhythmic pulse of these imitative encounters that increases in intensity to a point where any boundary between the brain and the sensory environment becomes indistinguishable. This is a folded intergenerational cycle of care and carelessness.

The rhythmic frequencies of the Tardean brain are endemic to a mode of neurocapitalism that not only requires attentive subjects but also synchronizes affective states. There is, evidently, a familiar Huxleyesque component to the quickening entrainment (and control) of the rhythmic brain's neurological energies engaged in joyful encounters. This is a rhythmic control of affective states perfectly captured in Huxley's hypnopedic rhymes, wherein the aesthetic power of repetitive drums, and harmonious chords, produces an inescapably haunting rhythm, capable of gathering attention and quelling any misguided thoughts of nonconformity. So what kind of alternative brain wave care systems might potentially resist the synchronization and assimilate a neurodiversity of brain frequency? A more promising avenue perhaps, in terms of understanding the potential of the rhythmic brain, follows Walter Freeman's contention that musical entrainment can be seen as a powerful tool to overcome solipsism because it helps to connect individuals to the overlapping sensory environments of others.[88] Here the discussion also returns to Tarde's social brain insofar as rhythmic entrainments of the crowd brain produce a loss of the sense of self with the added acknowledgment that trancelike states induced by collective entrainments can break down preexisting habits and beliefs. Barbara Ehrenreich has similarly discussed how collective festivities of Dionysian joy can produce an active resistance to repressive regimes.[89] The rhythmic repetition of rituals can, like this, reconnect the individual to the crowd, releasing him from the horror of Nietzsche's solitary existence. To be sure, Ehrenreich charts the endeavor by Calvinist capitalists to limit such festivities and suppress the emotional outpouring of unruly crowds. But times change; indeed, following Protevi, today, a significant contrast needs to be made between the pouvoir of the joyful encounter with capitalism and the puissance of riotous collective assembly. Questions still need to be addressed beyond this project concerning what kinds of festivities and music reconnect to the crowd and produce counterrhythmic entrainments that can offset, or even neutralize, the frequency-following disciplines, already there, in capitalism. At what speeds do the conservative tendencies of Wexler's brain, which destroy difference, and the progressive novelties of a nomadic thought that interferes with the normalization of rhythmic entrainments and neurotypical locationism oscillate? What is the difference between the dissonant deterritorializations of the crowd brain enthralled by Dionysian, ecstatic festivals and rituals and the territorialization of joyful encounters with Nazi joy? Is there a crowd music for a crowd brain? These are perhaps questions for another project.

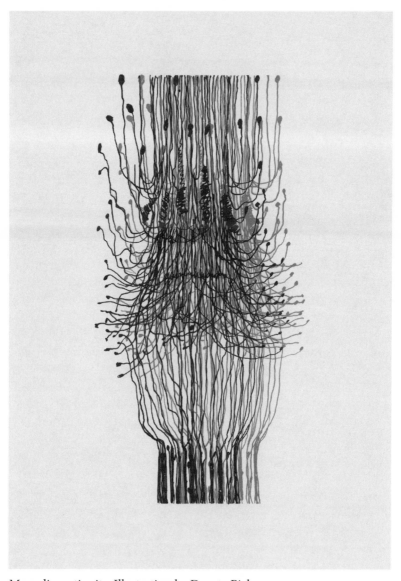

More discontinuity. Illustration by Dorota Piekorz.

CODA

Taking Care of the Not-Self

HUXLEY ESCAPES FROM PLATO'S CAVE

Perhaps it would be reckless for a theorist of critical interferences to attempt to emulate popular neuroscience writers by gazing into the future to look for answers to big philosophical questions, but here goes. Knowing what the potential energy of a brain wave will be in advance of it occurring is a complex thing to grasp. Although, as I have claimed, there is possibly more certitude in an oceanography of the brain than there is in the futurology of popular cognitive neuroscience, the biggest question of all perhaps for the brain sciences, concerning what constitutes consciousness, has not been answered, and neither will it be any time soon. The search for cognizant states, or, for that matter, nonconscious protosubjectivity, will certainly not lead us to the discovery of the essence of self. This is because the brain–somatic relation to the sensory environment is a multiplicity, an assemblage. As follows, subjectivity is always in the process of being made. So, on one hand, the mind–brain problem may never be solved. This is in part because the neurosciences face "fundamental barriers" when dealing with something as complex as the brain.[1] The brain *is* the chaos that continues to haunt science. This is a chaos that will not give up its secrets easily. This is a chaos that we encounter even though the black box has been opened up by neuroimaging technologies and physiological experimentation. What has been revealed is a further 100 billion impenetrable black boxes each with myriad hidden secrets. The diachronic and transversal emergence of each one of these black boxes ensures that the psychoneural equivalence between Cajal's butterflies of the soul and the sense of self will most probably remain an enigma for scientists, philosophers, and artists alike. This is because the multiplicity of parts in the brain are well and truly scrambled; the neuroimaging data are incomplete. On the other hand, the processes of the brain are intimately coupled to a radical relation to the outside. That is to say, the assemblage brain is, as Tarde contends, always-already *social*. Accordingly, we need to grasp everything the assemblages of sense making relate to, including chemical interactions, politicized sensory environments, and imitative encounters with caregivers. Without these looping social interactions, *without external stimulation*, there is, arguably, no sense.

More significantly, given the concerns of this book, the tension between the potential freedom of the assemblage brain and the potential dystopian future of neurocapitalism needs to be grasped. Although I have tried to neatly structure this book around two questions, there is perhaps a much messier *in-between* of *what can be done to a brain* and *what a brain can do*. If, for example, we return to Malabou's brain, which is already free but doesn't know it, we have to admit to a social brain that is caught between pouvoir and puissance. In other words, Tarde's somnambulist subject is, simultaneously, asleep and awake— *in-between* conscious states and a dream of action, *in-between* concept and sensation. What kind of freedom is this? How can we, in Malabou's terms, recue the uninhibited plasticity of the brain from its coincidence with the inhibited flexibilities of neurocapitalism, a mode of frequency-following capitalism that reaches out, via cultural circuits, to conform what remains of Gramsci's industrialized brain time? One way to approach this messiness is to engage in recent debates surrounding so-called neuroethics, which present a twisting vision of a future that is both, at once, Huxleyesque and yet full of nomadic potential. A good starting point for such a discussion on neuroethics is to perhaps follow Metzinger's endeavor to escape the cave brain through the exploration of altered states. Not only does this help him to escape the prison of his virtual self but it also draws attention to an ethics of altered states reminiscent of Huxley's soma-induced dystopia and the latter attempt, in *The Doors of Perception*, to use drugs to free, or *delimit*, the *Mind at Large*.

Resonant to Huxley's College of Emotional Engineering, Metzinger's Centre for Transpersonal Hedonic Engineering and Metaphysical Tunnel Design introduces a range of future neurotechnologies that "tinker with the hardware" of the cave brain.[2] These neurotechnologies take two forms. The first are technological tools designed to "tickle" the temporal lobe. "One can envisage," Metzinger contends, "a future in which people will no longer play video games or experiment with virtual reality for fun." These technologies, similar to some of the electrical brain stimulation products already available to us, in the form of EEG brain stimulation, would enable the exploration of altered states, so that we might discover consciousness outside the cave. The second type of neurotechnology is the psychoactive substance, which is again related to current cognitive enhancers, like Ritalin, Provigil, Adderall, and so on, but open up an imminent industry for "cosmetic psychopharmacy." This future will not only provide opportunities to augment cognition but allow for new freedoms to explore altered brain states and "design . . . conscious reality-models according to . . . wishes, needs and beliefs."[3]

Cosmetic psychopharmacy will evidently introduce a range of ethical problems for societies that continue to decree that some altered states of consciousness, like those attained from alcohol or prescription drugs, are legal, whereas others, like that from LSD, are not. Certainly, despite attempts by governments to control specific medicated altered states, the widespread use of the substances that attain such states is, according to Metzinger, inevitable. It is, after all, almost impossible to stop them. To begin with, the emergence of global Internet pharmacies poses significant new challenges to drug enforcement policies that already struggle to contain the flow of illegal drugs across physical borders. Illicit big pharma can also infinitely shift around the molecules of a drug to make illegal highs legal. Moreover, as the speed of life under capitalism increases, people will inexorably turn to cognitive enhancers to keep up with the demands to stay awake for longer hours, pay more attention, and remember more. Certainly these demands on brain time are necessary prerequisites for the speed rhythms of a digitally enhanced neurocapitalism. The market will provide a steady stream of products. So why should we be, as Metzinger terms it, neurophenomenological Luddites?[4]

It would seem that the onset of cosmetic psychopharmacy introduces an ethical dimension that simultaneously traverses the questions of *what can be done to a brain* and *what a brain can do,* to a point where any distinction between the two might in fact become nonsensical. Indeed, in the future, the gaps between the technologies that control brain time and those intended to make it free will become ever more indiscernible. The opening up of the neurological toolbox to every-body might therefore provide tools of control but also new weapons of subversion and resistance to efforts to conform the brain–somatic relation to the cadence of neurocapitalism. All the same, Metzinger's suggested neuroethics approach, which looks to determine what are good and bad states of consciousness, smacks of a Brave New World in which the end of the human suffering endured under the expansion of a virtual cybernetic brain state is achieved by way of what amounts to pedagogical brain training.[5]

OPIUM OR ACID?

Finally, it is Metzinger's direct references to Huxley that make for a con-clusive distinction between the cave brain and the assemblage brain. In short, Huxley's use of psilocybin, mescaline, and LSD in the late 1950s and early 1960s enabled him to grasp the brain-becoming-subject as a brain that limits itself. It is indeed Huxley who, after some thirty years of writing about dystopian brain control, graduates from soma to hal-lucinogens, marking something of a remarkable ethical about-face. He was now at the forefront of comprehending the transformational power of hallucinogenic drugs that might be able to remove filters that have previously blocked what he refers to as the *Mind at Large*. In this respect, Huxley was following Broad and Bergson's contention that the functions of the brain, the nervous system, and the sense organs were not productive but eliminative. To "make biological survival possible," Huxley remarks, "each one of us is potentially *Mind at Large*," which has to be "funnelled through the reducing valve of the brain and nervous system."[6] He continues,

> What comes out at the other end [of the funnel] is a measly trickle of the kind of consciousness which will help us to stay alive on the surface of this particular planet. To formulate and express the contents of this reduced awareness, man has invented and end-lessly elaborated those symbol-systems and implicit philosophies which we call languages. Every individual is at once the beneficiary and the victim of the linguistic tradition into which he has been born—the beneficiary inasmuch as language gives access to the

accumulated records of other people's experience, the victim in so far as it confirms him in the belief that reduced awareness is the only awareness and as it bedevils his sense of reality, so that he is all too apt to take his concepts for data, his words for actual things.[7]

As is the case with Deleuze and Guattari's chaos brain, Huxley's *Mind at Large* acts like an umbrella that protects us from being overwhelmed and confused by an unfiltered capacity to perceive and recollect everything that is happening everywhere in the universe. The brain blocks this overload, "leaving only that very small and special selection."[8]

Keeping with the drug theme, there is a significant distinction that needs to be made here between the kind of altered states Metzinger suggests free us from Plato's cave and Huxley's Bergson-inspired realization of uninhibited brain freedom. This is a difference that can be understood through a contrast made between two kinds of chemical experience and, as such, potentially counter brain rhythms to those experienced in neurocapitalism: first, Thomas De Quincey's opium eater, and second, Huxley's acid trip. On one hand, De Quincey's brain is trapped inside the cave. His consumption of opium is supposed to induce a mode of confessional dreaming, but this state of reverie needs to be mustered up from inside the prison of the self. Like this, opium enables De Quincey's brain to engage in a tendency toward an inner, isolated, and dreamy contemplation that is supposed to counter the pace of an external chaos quickened by the machines and crowds of increasingly industrialized urban social environment. It is, as such, the somnambulist *within* who, De Quincey contends, slows down the rush of rampant capitalism; otherwise, the "action of thoughts and feelings" will be "consciously dissipated and squandered" in these new rhythmic environments.[9] Nothing suffers more in the encounter with machines and crowds, it would seem, than the capacity of the brain to dream in solitude. As De Quincey notes,

> the machinery for dreaming planted in the human brain was not planted for nothing. That faculty, in alliance with the mystery of darkness, is the one great tube through which man communicates with the shadowy. And the dreaming organ, in connection with the heart, the eye, and the ear, compose the magnificent apparatus which forces the infinite into the chambers of a human brain, and

throws dark reflections from eternities below all life upon the mirrors of that mysterious camera obscura—the sleeping mind.[10]

It is De Quincey's opium habit that not only revealed to him the power of dreams but also exposed the inner faculty of a brain that dreams a solitary world of representation. This is a dream machine that protects the brain from the chaos of the horrendous objective reality of life in early capitalist cities and factories by escaping to an inner reality of subjective shadows. It is from within these shadows that it is possible to see the horror of capitalism with *eyes closed.*

In sharp contrast, Huxley's acid trip helps him see beyond the tendency to create these myopic inscapes in favor of an objective existence that sees with its eyes wide open. So why, he asks, is there this tendency in Western literature and philosophy to look *within* rather than *out there,* that is to say, to place a "spiritually higher significance" to seeing with eyes shut than it is with eyes open?[11] Huxley offers an explanation that interestingly reverses Wexler's notion that the brain prefers familiarity to difference. On the contrary, Huxley argues, "familiarity breeds contempt." We wake up to the external world "every morning of our lives."[12] This is the place where we experience the conformities of work. It is perhaps no surprise, then, that there is this tendency to escape to an inner world of "dreams and musings."[13] But, as Huxley goes on to argue, it is better to look beyond these inner visions, to stare out into the Void—"through the Void"—at the "the ten thousand things" of objective reality.[14] Even when Huxley did close his eyes during his mescaline-induced trips, he did not experience an inner world of dreams that could counter the monotonous repetitions of the external world of work. No other humans or animals appeared. He did not venture into strange landscapes with metamorphosing buildings or other magical spaces. Certainly he saw nothing remotely like a drama or a parable of the kind De Quincey encountered on opium—just an abstract world of gray structures and pale bluish spheres emerging into intense solidifications. Yet, when he opened his eyes, the mescaline admitted Huxley into another world, not of inner visions, but a world that existed *out there.* "The great change" he came across had nothing to do with his "subjective universe" but occurred in the "realm of objective fact," where, Huxley contended, the mescaline had exposed to him the "naked experience" of the external world.[15] This is not so much a brain that is, in the hippy vernacular, tuned in or turned on as it is a brain with its inhibitor turned off.

At breakfast, before Huxley had dropped his pill, he had noticed a vase containing three flowers. He had been struck by the "lively dissonance of its colors."[16] By eleven, though, he was not looking at the same vase. In his altered brain state, Huxley found himself neither inside Plato's cave nor, indeed, outside looking in. He was rather in the fold, where the inside is nothing but a fold of the outside. This naked experience of the vase, the *Istigkeit* or the is-ness of it all,[17] is a state of consciousness in which the sealed tunnels of the cave become invaded by the outside. It is this state of consciousness in which processes come into play, a becoming that Plato seems to have misinterpreted as a shadowy world of representations cast on the wall. As Huxley argues, Plato "made the enormous, the grotesque mistake of separating Being from becoming and identifying it with the mathematical abstraction of the Idea":[18]

> [Plato] could never, poor fellow, have seen a bunch of flowers shining with their own inner light and all but quivering under the pressure of the significance with which they were charged; could never have perceived that what rose and iris and carnation so intensely signified was nothing more, and nothing less, than what they were—a transience that was yet eternal life, a perpetual perishing that was at the same time pure Being, a bundle of minute, unique particulars in which, by some unspeakable and yet self-evident paradox, was to be seen the divine source of all existence.[19]

De Quincey's paranoid retreat into the cave is, like this, countered by Huxley, a philosopher king, *out of his head* (literally) on psychedelic philosophy. He is the schizoid acid freak who sees the clear potential of the assemblage brain, not to escape from the virtual world of dreams, but to free the brain and see the world for what it is. Huxley's brain becomes part of the assemblage: the *Not-Self.* It only knows an outside.

Acknowledgments

This book follows up on my previous effort for the University of Minnesota Press, *Virality,* insofar as it tries to answer some of the unanswered problems that cropped up in that book. It is also a product of various speaking engagements, discussions, and invitations to contribute to books and journals, stemming from its publication in 2012. I therefore need to thank all those people who invited me to discuss *Virality* and think through nascent ideas for *The Assemblage Brain.* Indeed, I have to apologize to many of these individuals for turning up with half-baked notions of what I was trying to achieve here. I hope this effort goes some way toward providing answers to the questions I attempted to sidestep at the time. So, in no particular order, thanks to all those who engaged with me, including Matt Fuller (Goldsmiths), Christian Borch (CBS), Steve Hinchliffe and Michael Schillmeier (Exeter), Sanjay Sharma (Brunel), the artists, Charlie Tweed (United Kingdom), and Alina Popa and Irina Gheorghe (Bureau of Melodramatic Research and all Bezna people in Bucharest), Obsolete Capitalism and Rhizomatika (Italy), Arthur and Marilouise Kroker (Ctheory), Joanne Garde-Hansen (Warwick), Erich Hörl (BKM), Tero Karppi and Mona Mannevuo (Buffalo and Turku), Carsten Stage (Aarhus), John Protevi (who kindly demonstrated entrained walking to me on the way to a seminar in Aarhus), Vincent Duclos and Vinh-Kim Nguyen (Paris and Montreal), Jussi Parikka and Yigit Soncul (WSA), Ana Gross and Nathaniel Tkacz (Warwick), Dave Boothroyd (Lincoln), and my co-organizers at Club Critical Theory in Southend-on-Sea for letting me occasionally talk about my work in a pub with a beer in hand.

Special thanks to Giles Tofield and Peter Vadden at the Cultural Engine in Southend-on-Sea for providing me with a great space to work during (and beyond) the final stages of writing the book. The endless supply of coffee and the occasional game of table football (all of which

I lost) helped me to get to this stage. Thanks also to all people at UEL who signed off on my sabbatical in 2014 and to Gosia Kwiatkowska, who covered classes for me.

Special gratitude goes to all of my family, particularly, on this occasion, Harry Sampson, who, at the very beginning of this project, informed me about the inspirational idea that if we were to remove the top of our skulls and reach inside to touch our brains, we'd feel nothing.

Finally, no thanks whatsoever go to the bloody Tory government in the United Kingdom, which continues to attack free public education as a human right. Shame on all you overprivileged, publicly educated sons and daughters of bankers who voted to cut maintenance grants for students from low-income families and turn them into debt! *No Ifs, No Buts, No Education Cuts!*

Notes

INTRODUCTION

1   Arthur Kroker, *Exits to the Posthuman Future* (Cambridge: Polity Press, 2014), 33.

2   Edmund T. Rolls, *Neuroculture: On the Implications of Brain Science* (Oxford: Oxford University Press, 2012).

3   Steven Rose, *The Future of the Brain: The Promise and Perils of Tomorrow's Neuroscience* (Oxford: Oxford University Press, 2005).

4   Murray Goldstein, "Decade of the Brain: An Agenda for the Nineties," *Western Journal of Medicine* 161, no. 3 (1994): 239–41.

5   University of South Carolina, "This Is Your Brain on Politics: Neuroscience Reveals Brain Differences between Republicans and Democrats," *Science Daily,* November 1, 2012, http://www.sciencedaily.com/releases /2012/11/121101105003.htm.

6   Rolls, *Neuroculture.*

7   Steve Harrison, Deborah Tatar, and Phoebe Sengers, "The Three Paradigms of HCI," paper presented at the Conference on Human Factors in Computing Systems, 2007.

8   Ibid.

9   Sally Satel and Scott O. Lilienfeld, *Brainwashed: The Seductive Appeal of Mindless Neuroscience* (New York: Basic Books, 2013).

10   Rose, *Future of the Brain,* 254–63.

11   A term borrowed from Nigel Thrift, *Knowing Capitalism* (London: Sage, 2005), 6.

12   See, e.g., Lisa Blackman, *Immaterial Bodies: Affect, Embodiment, Mediation* (London: Sage, 2012), and Blackman, "Affect and Automaticy: Towards an Analytics of Experimentation," *Subjectivity* 7 (2014): 362–84.

13   On one hand, nonrepresentational theory stands accused of only being interested in affect because it is not discourse. On the other hand, affect

is being inextricably linked to the guiding forces of meaning making, semiotics, and discourse. See Margaret Wetherell, *Affect and Emotion: A New Social Science Understanding* (London: Sage, 2012), 19.

14 Henri Bergson, *Matter and Memory* (London: George Allen and Unwin, 1911), 52–53.

15 Aside from steering away from the hyperbolized claims of popular neuroscience, I tend to agree with William R. Uttal that there are far too many barriers to overcome in terms of knowing what consciousness is, let alone what causes it. Uttal, *Neural Theories of the Mind: Why the Mind Brain Problem May Never Be Solved* (London: Routledge, 2014), 260–62.

16 As a recent example of this tension between, on one hand, idealism in the humanities and, on the other, posthumanism and the neurosciences, see Arthur Krystal, "The Shrinking World of Ideas," *Chronicle of Higher Education,* November 21, 2014, http://chronicle.com/article /Neuroscience-Is-Ruining-the/150141.

17 There is a flat sheet of cells that eventually folds in on itself to become the neural groove, then a tunnel, and a tube, in which three swellings form the brain. During this time, cells called neuroblasts with a capacity to divide will, over several months, grow into baby neurons, which in turn grow axons and dendrites and search for partners with which to make synapses. See Rose, *Future of the Brain,* 66–68.

18 The importance of which was pointed out to me by the artist and writer Alina Popa at the Bezna conference in Bucharest in June 2013 and then again in October 2015 at the Affect Theory: WTF conference in Lancaster, Massachusetts, where Steven Shaviro delivered a fascinating keynote on slime molds and decision making. See a similar discussion on Shaviro's *Pinocchio Theory* website called "Fruit Flies and Slime Molds," http://www.shaviro.com/Blog/?p=955.

19 Gilles Deleuze and Félix Guattari, *What Is Philosophy?* (London: Verso, 1994), 213.

20 As he writes in *Difference and Repetition* (London: Continuum, 2001), "we do not contemplate ourselves, but we exist only in contemplating— that is to say, *in contracting that from which we come*" (74, emphasis added). Indeed, and here we find an early version of what will, in *What Is Philosophy?,* become "micro-brains." For Deleuze, "everything is contemplation, even rocks and woods, animals and men . . . even our actions and needs" (75). See Gilles Deleuze, cited in Andrew Murphie,

"Deleuze, Guattari, and Neuroscience," in *The Force of the Virtual: Deleuze, Science and Philosophy,* edited by Peter Gaffney (Minneapolis: University of Minnesota Press, 2010), 283.

21 Deleuze and Guattari, *What Is Philosophy?,* 210.

22 "We read in *What Is Philosophy?* that 'the non philosophical is perhaps closer to the heart of philosophy than philosophy itself' (41). The negative, "non philosophical," does not designate any lack. It designates heterogeneity, positive divergence and contingent reason. It designates the need for an encounter that does not explain but produces—what Deleuze and Guattari called an heterogenesis, something new created in between two terms that keep their heterogeneity." Isabelle Stengers, "Gilles Deleuze's Last Message," 2006, http://www.recalcitrance.com /deleuzelast.htm. There is also another discussion to be had elsewhere with regard to François Laruelle's invention of non philosophy, which Deleuze and Guattari refer to as "one of the most interesting undertakings in philosophy." Deleuze and Guattari, *What Is Philosophy?,* 220.

23 Most of the references to Gabriel Tarde's social theory in this book are developments on my previous work. See Tony D. Sampson, *Virality: Contagion Theory in the Age of Networks* (Minneapolis: University of Minnesota Press, 2012).

24 Rolls, *Neuroculture,* 306.

25 V. S. Ramachandran, *The Tell-Tale Brain: Unlocking the Mystery of Human Nature* (London: Random House, 2011), 192–93.

26 Ibid.

27 Catherine Malabou, *What Should We Do with Our Brains?* (New York: Fordham University Press, 2008), 12.

28 Ibid.

29 Ibid., 11.

30 M. R. Bennett and P. M. S. Hacker, *History of Cognitive Neuroscience* (Chichester, U.K.: Wiley-Blackwell, 2013), 241, 243–44.

31 See Thomas Metzinger, *The Ego Tunnel: The Science of Mind and the Myth of the Self* (New York: Basic Books, 2009), and Gilles Deleuze, *The Fold: Leibniz and the Baroque* (Minneapolis: University of Minnesota Press, 1993).

32 Christian Borch, "Urban Imitations: Tarde's Sociology Revisited," *Theory, Culture, and Society* 22 (2005): 81–100.

33 John Protevi, "Deleuze and Wexler: Thinking Brain, Body, and Affect in Social Context," in *Cognitive Architecture: From Bio-politics to*

Noo-politics; Architecture and Mind in the Age of Communication and Information, ed. Deborah Hauptmann and Warren Neidich, 168–83 (Rotterdam, Netherlands: 010 Publishers, 2010).

1. INTERFERENCES

1   Bennett and Hacker, *History of Cognitive Neuroscience*, 250.

2   Ibid.

3   Gabe Turow and James D. Lane, "Binaural Beat Stimulation: Altering Vigilance and Mood States," in *Music, Science, and the Rhythmic Brain: Cultural and Clinical Implications*, ed. Jonathan Berger and Gabe Turow, 122–36 (New York: Routledge, 2011).

4   Arkady Plotnitsky, "From Resonance to Interference: The Architecture of Concepts and the Relationships among Philosophy, Art, and Science in Deleuze and Deleuze and Guattari," *Parallax* 18, no. 1: 22.

5   Ibid.

6   Turow and Lane, "Binaural Beat Stimulation," 122–36.

7   Deleuze and Guattari, *What Is Philosophy?*, 216–18.

8   Gilles Deleuze, "N is for Neurology," in *Gilles Deleuze from A to Z*, trans. Charles J. Stivale, dir. Pierre-André Boutang (Cambridge, Mass.: MIT Press, 2011), DVD.

9   Deleuze, cited in Murphie, "Deleuze, Guattari, and Neuroscience," 278.

10 Deleuze, "N is for Neurology."

11 Gilbert Cabasso and Fabrice Revault d'Allonnes, "On the Time-Image: A Conversation with Gilles Deleuze," in *Gilles Deleuze, Negotiations: 1972–1990* (New York: Columbia University Press, 1995), 60.

12 Ibid.

13 Gilles Deleuze, interview conducted by François Ewald and Raymond Bellour in *Magazine littéraire*, no. 257, September 1988.

14 Cabasso and Revault d'Allonnes, "On the Time-Image," 149.

15 Ramachandran, *Tell-Tale Brain*, 248.

16 Rosi Braidotti, "Elemental Complexity and Relational Vitality: The Relevance of Nomadic Thought for Contemporary Science," in Gaffney, *Force of the Virtual*, 212–14.

17 Ibid., 214.

18 Deleuze and Guattari, *What Is Philosophy?*, 209.

19 Ibid., 199, 209.

20 Bennett and Hacker, *History of Cognitive Neuroscience*, 103.

21 Ibid., 105–6.

22 Gilles Deleuze and Félix Guattari, *A Thousand Plateaus: Capitalism and Schizophrenia* (Minneapolis: University of Minnesota Press, 1987), 2.

23 Bennett and Hacker, *History of Cognitive Neuroscience*, 110–11.

24 Michael Schillmeier, *Eventful Bodies: The Cosmopolitics of Illness* (Farnham, U.K.: Ashgate, 2014), 57.

25 Ibid., 56.

26 Deleuze and Guattari, *A Thousand Plateaus*, 2.

27 Ibid., 2.

28 Ibid., 15–16.

29 "Deleuze and Guattari's 'posthumanism' is molecular biology." Murphie, "Deleuze, Guattari, and Neuroscience," 297.

30 Robert G. Shulman, *Brain Imaging: What It Can (and Cannot) Tell Us about Consciousness* (Oxford: Oxford University Press, 2013), 93.

31 Deleuze and Guattari, *What Is Philosophy?*, 149.

32 Tiziana Terranova, "Attention, Economy, and the Brain," *Culture Machine* 13 (2012), http://www.culturemachine.net/index.php/cm/article/view/465/484.

33 Ramachandran, *Tell-Tale Brain*, 117–35.

34 Kroker, *Exits to the Posthuman Future*, 40.

35 Ibid.

36 Braidotti, "Elemental Complexity and Relational Vitality," 214.

37 The scientific methods Deleuze and Guattari approached in the 1990s are radically different from those applied in the current wave of neuroscience. We need to therefore be aware of the problem of approaching neuroscience as a kind of *science in general* and take into account intensive approaches to science. See Manuel Delanda and Peter Gaffney, "The Metaphysics of Science: An Interview with Manuel Delanda," in Gaffney, *Force of the Virtual*, 329–32.

38 Adam Hadhazy, "Think Twice: How the Gut's 'Second Brain' Influences Mood and Well-Being: The Emerging and Surprising View of How the Enteric Nervous System in Our Bellies Goes Far beyond Just Processing the Food We Eat," *Scientific American*, February 12, 2010, http://www.scientificamerican.com/article.cfm?id=gut-second-brain.

39 Plotnitsky, "From Resonance to Interference," 27.

40 Deleuze and Guattari, *What Is Philosophy?*, 199.

41 Kroker, *Exits to the Posthuman Future*, 38.

42 Ibid.

43 Plotnitsky, "From Resonance to Interference," 22.

44 Ibid.

45 Deleuze and Guattari, *What Is Philosophy?*, 202.

46 Ibid., 118.

47 Ibid.

48 Ibid., 20.

49 Tony D. Sampson, "Tarde's Phantom Takes a Deadly Line of Flight—From Obama Girl to the Assassination of Bin Laden," *Distinktion* 13, no. 3 (2012): 358–59.

50 Peter Gaffney, "Introduction: Science in the Gap," in Gaffney, *Force of the Virtual*, 2–7.

51 Ibid., 5.

52 Ibid., 6–7.

53 Deleuze and Guattari, *What Is Philosophy?*, 126.

54 Ibid., 127.

55 Ibid., 212.

56 Ibid., 177.

57 Ibid., 198.

58 For a more detailed discussion on Deleuze and Guattari and Duchamp's ready-mades, see Stephen Zepke, *Art as Abstract Machine: Ontology and Aesthetics in Deleuze and Guattari* (New York: Routledge, 2005), 159–64.

59 Ibid., 125.

60 As is supposed to be the case in electrical synaptic transmissions. See Joseph LeDoux, *The Synaptic Self: How Our Brains Become Who We Are* (New York: Penguin Books, 2003), 61.

61 Deleuze and Guattari, *What Is Philosophy?*, 198.

62 Ibid.

63 See Stephen Zepke, "The Post-conceptual Is the Non-conceptual: Deleuze and Guattari and Conditions of Contemporary Art," 2009, http://www.actualvirtualjournal.com/2014/11/the-post-conceptual-is-non-conceptual.html.

64 Gilles Deleuze, "Postscript on the Societies of Control," in *Rethinking Architecture: A Reader in Cultural Theory,* ed. Neil Leach (New York: Routledge, 1997), 295.

65 Charles Saatchi, "Charles Saatchi: The Hideousness of the Art World: Even a Show-Off Like Me Finds This New, Super-rich Art-Buying Crowd Vulgar and Depressingly Shallow," *The Guardian,* December 2011, http://www.theguardian.com/commentisfree/2011/dec/02/saatchi-hideousness-art-world.

66 From a discussion between Ricardo Basbaum and the author in London, July 2013.
67 Deleuze and Guattari, *What Is Philosophy?*, 217.
68 Ibid., 128–29.
69 Ibid., 66–63. Borrowing their "new" idiot from Dostoyevsky.
70 Ibid., 63.
71 Schillmeier, *Eventful Bodies*, 43.
72 Deleuze and Guattari, *What Is Philosophy?*, 177.
73 Ibid., 66.
74 Ibid., 177.
75 Ibid., 127.
76 Ibid., 132.
77 Ibid.
78 Ibid., 175.
79 Aldous Huxley, *The Doors of Perception*, https://www.maps.org/images /pdf/books/HuxleyA1954TheDoorsOfPerception.pdf.
80 Ibid., 165. Indeed, the conceptualization of drugs, we are told, makes them "extraordinarily flaky, unable to preserve themselves."
81 Ibid., 35.
82 Ibid., 202.
83 Ibid., 203.
84 Ibid., 192–93.
85 Ibid., 188.
86 Ibid., 66.
87 Ibid., 118.
88 Ibid., 123.
89 Ibid., 122.
90 Ibid., 124.
91 Ibid., 119.
92 Deleuze and Guattari, *A Thousand Plateaus*, 367.
93 Ibid.
94 Ibid., 486.
95 Deleuze and Guattari, *What Is Philosophy?*, 161.
96 Isabelle Stengers, "Experimenting with *What Is Philosophy?*," in *Deleuzian Intersections: Science, Technology, Anthropology*, ed. Casper Bruun Jensen and Kjetil Rödje, 39–56 (New York: Berghahn Books, 2010).
97 Ibid., 39.
98 Ibid., 52–53.

99   François Dosse, *Gilles Deleuze and Félix Guattari: Intersecting Lives* (New York: Columbia University Press, 2010), 15.
100 Stengers, "Experimenting with *What Is Philosophy?*," 53.
101 Ibid., 41.
102 Deleuze and Guattari, *A Thousand Plateaus,* 161.
103 Deleuze and Guattari, *What Is Philosophy?*, 218.
104 Ibid., 66.
105 Ibid., 101.
106 Ibid., 109.
107 Stengers, "Gilles Deleuze's Last Message."
108 Deleuze and Guattari, *What Is Philosophy?*, 218.
109 Arkady Plotnitsky, "Images of Thought and the Sciences of the Brain," in Gaffney, *Force of the Virtual,* 258–59.
110 Eric Alliez, *The Signature of the World, or, What Is Deleuze and Guattari's Philosophy?* (New York: Continuum, 2004), 51.

PART I

1   J. Javier Campos-Bueno and Antonio Martín-Araguz, "Neuron Doctrine and Conditional Reflexes at the XIV International Medical Congress of Madrid of 1903," *Psychologia Latina* 3, no. 1 (2012): 10–22, http://www.ucm.es/info/psyhisp/es/5/art24.pdf.
2   Ibid., 17.
3   Ibid., 15.
4   Ibid.
5   It would, however, be another fifty years before the invention of electron microscopic technologies that could actually observe Sherrington's synapse with the invention of the electron microscope.
6   Javier Bandrés and Rafael Llavona, "Pavlov in Spain," *Spanish Journal of Psychology* 6, no. 2 (2003): 81–92.
7   Ibid., 82.
8   Ibid.
9   Ibid.
10 Rose, *Future of the Brain,* 241.
11 Ibid.
12 Ben Ehrlich, "A Portrait of the Scientist as a Young Artist," July 1, 2010, http://www.thebeautifulbrain.com/pdf/Cajal_Portrait_Ehrlich.pdf.

13 Cajal's father apparently considered his obsession for art a developmental defect. Santiago Ramón y Cajal, *Recollections of My Life* (Cambridge, Mass.: MIT Press, 1989), 48–49.

14 Sally Weale, "ADHD Drugs Increasingly Prescribed to Treat Hyperactivity in Pre-schoolers: A Fifth of Educational Psychologists Say They Know of Children Being Given Medication Despite Guidelines Advising Against It," *The Guardian*, December 21, 2014, http://www.theguar dian.com/society/2014/dec/21/adhd-medication-treat-hyperactivity-pre -school-children.

15 Cajal, *Recollections of My Life*, 42.

16 W. M. Cowan, foreword to Cajal, *Recollections of My Life*, vii.

17 Santiago Ramón y Cajal, "The Fabricator of Honor," in *Vacation Stories: Five Science Fiction Stories* (Urbana, Ill.: University of Illinois Press, 2001), 67.

18 Ibid., 68.

19 Ibid., 51.

2. NEUROLABOR

1 Although it must be noted that authors in HCI do engagement with politics. See, for example, Phoebe Sengers, "The Ideology of Modernism in HCI," http://www.cl.cam.ac.uk/events/experiencingcriticaltheory /Sengers-IdeologyModernism.pdf.

2 Tiziana Terranova, "Free Labor: Producing Culture for the Digital Economy," *Social Text* 63, vol. 18, no. 2 (2000): 33–58.

3 For a more detailed account of the subtleties involved, see Tiziana Terranova, "Debt and Autonomy: Lazzarato and the Constituent Powers of the Social," *The New Reader* 1 (2014), http://thenewreader.org/Issues /1/DebtAndAutonomy.

4 Ibid.

5 Antonio Gramsci, in David Forgacs, ed., *The Gramsci Reader: Selected Writings 1916–1935* (New York: New York University Press, 2000), 295.

6 Ibid.

7 Ibid.

8 Ibid.

9 Henry Ford, *My Life and Work* (Minneapolis: Filiquarian, 2006), 52.

10 Ibid., 310.

11   See Michael Dieter, "Contingent Operations: Transduction, Reticular Aesthetics, and the EKMRZ Trilogy," in *Error: Glitch, Noise, and Jam in New Media Cultures,* ed. Mark Nunes (New York: Continuum, 2011), 191.

12   See, e.g., Reut Schwartz-Hebron, "Using Neuroscience to Effect Change in the Workplace," *Employment Relations Today* 39, no. 2 (2012): 11–15; Steven Kotler, "Corporate Communication: A Prominent Neuroscientist's Take on the Subtle Ninjitsu of Workplace Conversation," *Forbes,* July 4, 2012, http://www.forbes.com/sites/stevenkotler/2012/07/24/corporate-communication-a-prominent-neuroscientists-take-on-the-subtle-nin jitsu-of-workplace-conversation/; and Erika Garms, *The Brain Friendly Workplace: 5 Big Ideas from Neuroscience That Address Organizational Challenges* (Alexandria, Va.: ASTD Press, 2014).

13   Bernard Stiegler, "From Neuropower to Noopolitics," paper presented at the Unlike Us Conference, Institute of Network Cultures, Amsterdam, March 22, 2013, https://vimeo.com/channels/unlikeus3/63803603. See also Bernard Stiegler, *Taking Care of Youth and the Generations* (Stanford, Calif.: Stanford University Press, 2010), 124–29.

14   Joseph Pine and James H. Gilmore, *The Experience Economy,* updated ed. (Boston: Harvard Business School Press, 2011).

15   Ibid., 17–24.

16   See, e.g., Don Norman, *Emotional Design: Why We Love (or Hate) Everyday Things* (New York: Basic Books, 2005), and Susan M. Weinschenk, *Neuro Web Design: What Makes Them Click?* (Berkeley: New Riders, 2008).

17   This work is influenced by Harrison et al., "Three Paradigms of HCI," a text Matthew Fuller referred to when commenting on my "third paradigm" research at a seminar called *Studies in Evil Media* at the Centre for Cultural Studies Research, University of East London, on October 7, 2009. Fuller also pointed me toward Brigitte Kaltenbacher's excellent PhD thesis "Intuitive Interaction: Steps towards an Integral Understanding of the User Experience in Interaction," Goldsmiths, University of London.

18   Harrison et al., "Three Paradigms of HCI."

19   See, e.g., Norman, *Emotional Design,* 12.

20   Thomas Kuhn, *The Structure of Scientific Revolutions* (Chicago: University of Chicago Press, 1996).

21   Thrift, *Knowing Capitalism,* 93–94.

22   It was not, however, until 2009 that the Ergonomic Society in the United Kingdom was renamed the Chartered Institute of Ergonomics and

Human Factors (IEHF), reflecting the popular usage of both terms and to emphasize the breadth of the discipline. See the IEHF website at http://www.ergonomics.org.uk/about-us/history/.

23 Harrison et al., "Three Paradigms of HCI."

24 Jonathan Crary, *Suspensions of Perception: Attention, Spectacle, and Modern Culture* (London: MIT Press, 2001).

25 Jenny Preece, *Human–Computer Interaction* (Wokingham, U.K.: Addison Wesley, 1994), 101.

26 It is significant to note that stress is now the single biggest cause of absence from work in the United Kingdom, having overtaken repetitive strain injury in 2012. See William Davies, "John Lewis and a New Vision for Capitalism: It Is Time for Companies to Become Something Other Than Just Vehicles for Making Money," *Daily Telegraph,* January 15, 2012, http://www.telegraph.co.uk/finance/economics/9014227/John -Lewis-and-a-new-vision-for-capitalism.html.

27 Bannon, as cited in Preece, *Human–Computer Interaction,* 69.

28 BBC News, "Amazon Workers Face 'Increased Risk of Mental Illness,'" November 25, 2013, http://www.bbc.co.uk/news/business-25034598.

29 John McCarthy and Peter Wright, *Technology as Experience* (Cambridge, Mass.: MIT Press, 2007), 12.

30 Harrison et al., "Three Paradigms of HCI."

31 Adam Greenfield, *Everyware: The Dawning of the Age of Ubiquitous Computing* (Berkeley: New Riders, 2006), 2.

32 Michael Millar, "Union Calls for Halt to RFID Tracking of Workers," *Personnel Today,* July 18, 2005, http://www.personneltoday.com/hr/union -calls-for-halt-to-rfid-tracking-of-workers/.

33 Harrison et al., "Three Paradigms of HCI."

34 Kirsten Boehner, Rogério DePaula, Paul Dourish, and Phoebe Senger, "Affect: From Information to Interaction," in *Proceedings on Critical Computing,* 59–68 (New York: ACM Press, 2005), http://citeseerx.ist .psu.edu/viewdoc/download?doi=10.1.1.86.6303&rep=rep1&type=pdf.

35 Harrison et al., "Three Paradigms of HCI."

36 In some cases making an important distinction between Damasio's emotional thesis and the cultural interpretation of emotional interactions. See Rogério DePaula and Paul Dourish, "Cognitive and Cultural Views of Emotions," paper presented at the Human Computer Interaction Consortium Winter Meeting, 2005, http://www.dourish.com /publications/2005/hcic2005-emotions.pdf.

37 Damasio, as cited in Tony D. Sampson, "Contagion Theory beyond the Microbe," in *Critical Digital Studies: A Reader*, 2nd ed., ed. Arthur Kroker and Marilouise Kroker (Toronto: University of Toronto Press, 2013), 124.

38 Norman, *Emotional Design*, 12.

39 Thrift, *Knowing Capitalism*, 6–7.

40 Weinschenk, *Neuro Web Design*, 65.

41 A. K. Pradeep, "The See-Through Consumer," *The Economist*, April 16, 2012, https://www.youtube.com/watch?v=-AkVX25adE0.

42 Nigel Thrift, "Remembering the Technological Unconscious by Fore-grounding Knowledges of Position," *Environment and Planning: Society and Space* 22 (2004): 175–90.

43 As initially developed in Sampson, *Virality*.

44 Nigel Thrift, *Non-representational Theory: Space/Politics/Affect* (London: Routledge, 2008), 32.

45 Pine and Gilmore, *Experience Economy*. See also B. Joseph Pine II and James H. Gilmore, "Welcome to the Experience Economy," *Harvard Business Review*, July–August 1998, https://hbr.org/1998/07/welcome -to-the-experience-economy.

46 Jussi Parikka, "'Tarde as Media Theorist': An Interview with Tony D. Sampson," *Theory, Culture, and Society*, January 25, 2013, http://theory culturesociety.org/tarde-as-media-theorist-an-interview-with-tony-d -sampson-by-jussi-parikka/.

47 Jesse Schell, "When Games Invade Real Life," talk presented at the DICE Summit, 2010, http://www.ted.com/talks/jesse_schell_when_games_in vade_real_life.

48 Ibid.

49 Greenfield, *Everyware*, 26.

50 Stiegler, *Taking Care of Youth and the Generations*, 100.

51 Ibid., 33.

3. CONTROL AND DYSTOPIA

1 Deleuze, "Postscript on the Societies of Control," 295.

2 Stuart Elliot, "Is the Ad a Success? The Brain Waves Tell All," *New York Times*, March 31, 2008, http://www.nytimes.com/2008/03/31/business /media/31adcol.html?_r=0.

3   Thomas Davenport and John Beck, *The Attention Economy: Understanding the New Currency of Business* (Boston: Harvard Business School Press, 2001).

4   Elliot, "Is the Ad a Success?"

5   See, e.g., Robert Heath, *Seducing the Subconscious: The Psychology of Emotional Influence in Advertising* (Chichester, U.K.: Wiley-Blackwell, 2012).

6   As indeed was mentioned at a marketing seminar attended by the author in 2013 in London.

7   Deleuze, "Postscript on the Societies of Control," 296.

8   Matthew Fuller and Andrew Goffey, "On the Usefulness of Anxiety: Two Evil Media Stratagems," *The Sari Reader 08: Fear* (Deli: Centre for the Study of Developing Societies, 2010), 156, http://archive.sarai.net /files/original/7f19cb086dd3c08c4860ddf48b61b0d2.pdf.

9   Ibid., 160.

10   Raymond Williams, *Culture and Society: Coleridge to Orwell* (London: Hogarth Press, 1987), 286.

11   Fordlândia is the so-called forgotten jungle city Henry Ford built in the heart of the Brazilian Amazon in the 1920s as a way to control the global production of rubber and replicate North American capitalism in South America. Fordlândia was a cross between a rubber plantation and a socially regulated North American Midwest factory town. It was the archetypal industrialist control society. Workers were provided with modern health care, education, and a golf course but were also subjected to a totalitarian state with strict prohibition laws, dietary regulations, and tightly controlled worker leisure time, which was put under surveillance. When synthetic rubber became available in the 1940s, Ford simply walked away from the project. Fordlândia is now a ruin in the jungle.

12   Gregory Corso and Allen Ginsberg, "Interview with William S. Burroughs," *Journal for the Protection of All People* (1961), http://realitystudio .org/interviews/1961-interview-with-william-s-burroughs-by-gregory -corso-and-allen-ginsberg/.

13   Ibid., 294.

14   With the possible exception of Burroughs's William Lee.

15   William E. Connolly, *Neuropolitics: Thinking, Culture, Speed* (Minneapolis: University of Minnesota Press, 2002), 51.

16   William Burroughs, "The Limits of Control," in *The Adding Machine: Collected Essays* (London: John Calder, 1985).

17 Aldous Huxley, *Brave New World, London* (1932; repr., London: Vintage, 2007), 42.

18 Ibid., 46.

19 Ibid.

20 Ibid.

21 Burroughs, "Limits of Control."

22 Stiegler, *Taking Care of Youth and the Generations.*

23 Care Quality Commission, "The Safer Management of Controlled Drugs," 2012 Annual Report, August 2013, http://www.cqc.org.uk/sites /default/files/documents/cdar_2012.pdf. It is possible that because many of these smart drugs are acquired through online pharmacies, the increases recorded by the CQC report might in fact be fairly modest.

24 Crary, *Suspensions of Perception.*

25 Ibid., 49.

26 Ibid., 35–37.

27 Ibid., 36.

28 Although the dialectic is dismissed, the reader will note that cultural pessimism remains intact.

29 A. K. Pradeep, "The Core of NeuroMarketing: Notice/Like/Remember," http://neurofocus.com.co/www/ceo_spotlight.htm. See also A. K. Pradeep, *The Buying Brain* (Hoboken, N.J.: John Wiley, 2010).

30 Stiegler, *Taking Care of Youth and the Generations,* 4–8.

31 Ibid., 5, emphasis added.

32 Ibid., 56.

33 Ibid.

34 Stiegler, "From Neuropower to Noopolitics."

35 See an elaboration on this point in Sampson, *Virality.*

36 Gabriel Tarde, "L'Opinion et la foule," in *Gabriel Tarde: On Communication and Social Influence,* ed. T. N. Clark, 277–96 (1901; repr., Chicago: Chicago University Press, 2010).

37 Aldous Huxley, "The Ultimate Revolution," talk presented at the University of California, Berkeley, March 20, 1962, http://sagaciousnews network.com/aldous-huxley-speech-at-uc-berkeley-the-ultimate-revo lution-1962/. See also Aldous Huxley, *Brave New World Revisited* (1958; repr., London: Vintage, 2004).

38 Despite continued uncertainty concerning what causes electrical activity in the brain, EEG recordings are largely attributed to postsynaptic potentials in cell bodies and dendrites, leading to the eventual firing of neurons. See Berger and Turow, *Music, Science, and the Rhythmic Brain,* 16.

39 In his 1962 speech at UC Berkeley, Huxley cited Grey Walter's account of the depressions of asylum inmates being controlled by electrodes.

40 Ibid.

41 Berger and Turow, *Music, Science, and the Rhythmic Brain,* 17.

42 Ibid., 16.

43 Ibid., 16, 19.

44 See, e.g., Giovanni Vecchiato, Laura Astolfi, Fabrizio De Vico Fallani, Jlenia Toppi, Fabio Aloise, Francesco Bez, Daming Wei, Wanzeng Kong, Jounging Dai, Febo Cincotti, Donatella Mattia, and Fabio Babiloni, "On the Use of EEG or MEG Brain Imaging Tools in Neuromarketing," *Computational Intelligence and Neuroscience* 2011 (2011), doi:10.1155/2011/643489.

45 The test, called the Neuropsychiatric EEG-Based Assessment Aid (NEBA) System, was approved for marketing by the U.S. Food and Drug Administration in July 2013. See the FDA website at http://www.fda.gov/newsevents/newsroom/pressannouncements/ucm360811.htm.

46 Rose, *Future of the Brain,* 256.

47 E. P. Thompson, "Time, Work-Discipline, and Industrial Capitalism," *Past and Present* 38 (1967): 56–97.

48 Nigel Thrift, *Spatial Formations* (London: Sage, 1996), 209–10.

49 Paul Virilio, *Speed and Politics: An Essay on Dromology* (New York: Semiotext(e), 1986).

50 Huxley, *Brave New World Revisited,* 91.

51 Rose, *Future of the Brain,* 7, 241.

52 Crary, *Suspensions of Perception,* 36.

53 Ibid.

54 LeDoux, *Synaptic Self,* 246–50.

55 Rolls, *Neuroculture,* 294.

56 LeDoux, *Synaptic Self.*

57 Rose, *Future of the Brain,* 258.

58 See, e.g., Jay Gunkelman, "Drug Exposure and EEG/qEEG Findings," Quantitative Electroencephalography (qEEG): Information and Discussion, 2009, http://qeegsupport.com/drug-exposure-and-eegqeeg-findings/.

59 Rose, *Future of the Brain,* 259.

60 Ibid., 260.

61 For example, dopamine is assumed to reward but also seems to serve many other functions involved in the control of locomotion, cognition, emotion, and affect as well as neuroendocrine secretion (the interactions

between the nervous system, hormones, and the glands). See "How Ritalin Works in Brain to Boost Cognition, Focus Attention," *Science Daily,* June 25, 2008, https://www.sciencedaily.com/releases/2008/06/080624115956.htm.

62   Psychiatrist David Healy, as cited in Rose, *Future of the Brain,* 260.

63   Linda Carroll, "'Steroids for School': College Students Get Hooked on 'Smart Drugs'—Hidden-Camera Investigation Shows How Easy It Is to Buy Illegal Prescription Drugs," *Today,* 2011, http://www.today.com/id/43050779/ns/today-today_health/t/steroids-school-college-students-get-hooked-smart-drugs/#.Ujl_iX9Bw4o.

64   Ibid., emphasis added.

65   Pietter Lemmens, "This System Does Not Produce Pleasure Anymore: An Interview with Bernard Stiegler," *Krisis: Journal for Contemporary Philosophy* 1 (2011): 33–41, http://www.krisis.eu/content/2011-1/krisis-2011-1-05-lemmens.pdf.

66   Ibid., 40.

67   Huxley, *Brave New World,* 21.

68   Both institutions were housed in a single sixty-story building on Fleet Street, the center for print, radio, and television production. The *Gamma Gazette,* the Hourly Radio, and Feeling Pictures were among the media outlets using the building, but the top eighteen floors were occupied by CEE.

69   Nicholas Carr, "The Manipulators: Facebook's Social Engineering Project," *Los Angeles Review of Books,* September 14, 2014, http://lareviewofbooks.org/essay/manipulators-facebooks-social-engineering-project#.

70   Adam D. I. Kramer, Jamie E. Guillory, and Jeffrey T. Hancock, "Experimental Evidence of Massive-Scale Emotional Contagion through Social Networks," *Proceedings of the National Academy of Sciences of the United States of America* 111, no. 24 (2014): 8788–90, http://www.pnas.org/content/111/24/8788.full.

71   Carr, "Manipulators."

72   Ibid.

73   Ibid.

74   Heath, *Seducing the Subconscious,* 15–16.

75   Edward Bernays, *Propaganda* (1928; repr., New York: IG, 2005), 77.

76   LeDoux, *Synaptic Self,* 120–24.

77   The neuroeconomist observes the functioning of dopamine during decision-making processes and generally maps them according to the

dopaminergic reward prediction error (DRPE) hypothesis, that is to say, a computational model of a neurotransmitter that releases dopamine in proportion to the difference between the "predicted reward" and the "experienced reward" of a particular event. Andrew Caplin and Mark Dean, "Dopamine, Reward Prediction Error, and Economics," 2007, http://cess.nyu.edu/caplin/wp-content/uploads/2010/02/Dopamine -Reward-Prediction-Error-and-Economics.pdf.

78 Ibid.

79 See, e.g., Damasio's keynote appearance at the Neuromarketing World Forum 2014 conference, http://www.neuromarketingworldforum.com /2014-new-york/speakers.

80 "ADHD: Making the Invisible Visible," a white paper facilitated and funded by Shire AG and supported by the European Brain Council and Global Alliance of Mental Illness Advocacy Networks, 2013, http://www .russellbarkley.org/factsheets/ADHD_MakingTheInvisibleVisible.pdf.

81 Schillmeier, *Eventful Bodies*, 12.

82 "ADHD: Making the Invisible Visible," 2.

83 Weale, "ADHD Drugs Increasingly Prescribed."

84 Fuller and Goffey, "On the Usefulness of Anxiety."

85 Deleuze, "Postscript on the Societies of Control," 293.

PART II

1 Joseph Needham, review of *Scrutiny* in Donald Watt, ed., *Aldous Huxley: The Critical Heritage*, 202–5 (New York: Routledge, 2013).

2 Ibid., 210–12.

3 Ibid., 205.

4 See Watt's introduction to *Aldous Huxley*, 16.

5 Ibid., 25. Joyce Cary cited in ibid.

6 Malabou, *What Should We Do with Our Brains?*

7 Ibid., 8.

8 Ibid.

9 Ibid.

10 Ibid., 11.

11 Huxley, *Brave New World*, 208

12 Wilhelm Reich, *The Mass Psychology of Fascism* (New York: Orgone Institute Press, 1946).

13 As previously expanded on in Sampson, *Virality*.

14 Maurizio Lazzarato, as cited in Thrift, *Non-representational Theory*, 230.

15 Gabriel Tarde, preface to the second edition of *The Laws of Imitation* (New York: Henry Holt, 1903), xi.

16 Tarde, as cited in Sampson, *Virality*, 92.

17 Ibid., 180. A comparison is made between George Lakoff's neurologically inspired political unconscious and Tarde's somnambulist.

18 Bruce E. Wexler, *Brain and Culture: Neurobiology, Ideology, and Social Change* (Cambridge, Mass.: MIT Press, 2008), 117–19.

19 Ibid., 115.

20 A comparison between Tarde's and Deleuze and Guattari's references to birdsong is developed in Sampson, "Tarde's Phantom Takes a Deadly Line of Flight," 358–59.

21 Ibid., 358.

22 Bergson, *Matter and Memory*, 51.

23 Gabriel Tarde, *Penal Philosophy* (1921; repr., New Jersey: Transaction, 2001), 116–18.

24 Deleuze, *The Fold*.

25 Water Lippman, *Public Opinion* (1922; repr., New York: Free Press Paperbacks, 1997).

26 Tarde, *Penal Philosophy*, 195.

4. SENSE MAKING AND ASSEMBLAGES

1 See, e.g., Pierre Lévy, *Collective Intelligence: Mankind's Emerging World in Cyberspace* (Cambridge: Perseus, 1997).

2 Gary Genosko, *When Technocultures Collide* (Waterloo, Ont.: Wilfrid Laurier University Press, 2013), 29.

3 See http://milosrajkovic.tumblr.com/post/52301260775/by-sholim.

4 Pierre Lévy, *Cyberculture* (Minneapolis: University of Minnesota Press, 2001), 229–30.

5 In the sense that the influence of Durkheim can be traced to theories of social emergence evident in cybernetics and systems theory. See Sampson, *Virality*, 34–35.

6 Kroker, *Exits to the Posthuman Future*, 37.

7 Pierre Lévy, *Becoming Virtual: Reality in the Digital Age* (New York: Plenum Trade, 1998).

8 Ibid., 9.

9   Pierre Lévy, *The Semantic Sphere 1: Computation, Cognition, and Information Economy* (London: John Wiley, 2011), 115.

10   Lévy, *Cyberculture*, 229–30.

11   Kroker, *Exits to the Posthuman Future*, 38.

12   Genosko, *When Technocultures Collide*, 29.

13   Ibid., 30–39. Genosko points out that big-toe computing has a history dating backing to the 1950s.

14   Deleuze and Guattari, *What Is Philosophy?*, 210–11.

15   LeDoux, *Synaptic Self*, 18.

16   Bergson, *Matter and Memory*, 12.

17   LeDoux, *Synaptic Self*, 32.

18   Ibid., 22.

19   Ibid., 120.

20   Antonio Damasio, *The Feeling of What Happens: Body, Emotion, and the Making of Consciousness* (London: Vintage, 2000), 21.

21   Ibid., 22.

22   Ibid.

23   Ibid., 153.

24   Ibid., 174.

25   LeDoux, *Synaptic Self*, 31.

26   Ibid., 29.

27   Malabou, *What Should We Do with Our Brains?*, 8.

28   Ibid.

29   Ibid., 58.

30   Ibid.

31   Ibid., 59–60, emphasis added.

32   Bergson, *Matter and Memory*, 9.

33   Ibid., 29.

34   Ibid., 9.

35   Ibid., 13.

36   Ibid., 208.

37   Ibid.

38   Ibid., 276.

39   Ibid., 276–77.

40   Ibid., 277.

41   Ibid., 209.

42   Ibid., 13.

43   Ibid., xvii.

44   Ibid., xvii–xviii.

45　Malabou, *What Should We Do with Our Brains?*, 3–5.

46　Ibid., 4–5.

47　Damasio, *Feeling of What Happens*, 317–23.

48　Ibid., 191–92.

49　Ibid., 192.

50　LeDoux, *Synaptic Self*, 76.

51　Ibid., 42.

52　Ibid., 76–77.

53　Manuel Delanda, *Philosophy and Simulation: The Emergence of Synthetic Reason* (London: Bloomsbury, 2013), 90–91.

54　Ibid., 90.

55　Ibid.

56　Malabou, *What Should We Do with Our Brains?*, 10.

57　Ibid., 12.

58　Ibid., 5.

59　Ibid., 12.

60　Ibid.

61　Ibid., 6.

62　Ibid.

63　Delanda, *Philosophy and Simulation*, 192.

64　Malabou, *What Should We Do with Our Brains?*, 71.

65　Ibid.

66　Ibid.

67　Ibid., 74.

68　Damasio, as cited in ibid.

69　Ibid., 74–75.

70　Gilles Deleuze, *Bergsonism* (1966; repr., New York: Zone Books, 2002), 42.

71　Metzinger, *Ego Tunnel*, 111.

72　Thomas Metzinger, *Being No One: The Self-Model Theory of Subjectivity* (Cambridge, Mass.: MIT Press, 2004), 547–51.

73　Nietzsche, as cited in Gilles Deleuze, *The Logic of Sense* (1969; repr., London: Continuum, 2004), 300.

74　Bennett and Hacker, *History of Cognitive Neuroscience*, 113.

75　Ibid.

76　Ibid., 240–63.

77　Ibid., 240.

78　Ludwig Wittgenstein, *The Collected Works of Ludwig Wittgenstein: The Blue and Brown Books* (Oxford: Blackwell, 1998), 7.

79   Ibid., 8.

80   Ibid., 6.

81   Bennett and Hacker, *History of Cognitive Neuroscience*, 241–45.

82   Wittgenstein, *Collected Works*, 8–9.

83   Deleuze and Guattari, *What Is Philosophy?*, 209–13.

84   Gilles Deleuze, "W is for Wittgenstein," in *Gilles Deleuze from A to Z*, emphasis added.

85   Schillmeier, *Eventful Bodies*, 56–57.

86   Darren Ellis and Ian Tucker, *The Social Psychology of Emotion* (London: Sage, 2015), 57.

87   See Steve D. Brown and Paula Reavy, *Vital Memory: Ethics, Affect, and Agency* (New York: Routledge, 2014).

88   Ibid., 57.

89   Ibid., 56.

90   Gilles Deleuze and Félix Guattari, *Anti-Oedipus* (London: Athlone Press, 1984), 40.

91   Ibid., 42.

92   Ibid.

93   Ibid., 42–43.

94   Delanda, *Philosophy and Simulation*, 3.

95   Ibid., 194.

96   Ibid., 184.

97   Manuel Delanda, *A New Philosophy of Society: Assemblage Theory and Social Complexity* (London: Continuum, 2006), 8–25.

98   Delanda, *Philosophy and Simulation*, 194.

99   John Protevi, "Deleuze, Guattari, and Emergence," *Paragraph: A Journal of Modern Critical Theory* 29, no. 2 (2006): 19–39.

100   Schillmeier, *Eventful Bodies*, 62, 74.

101   As established in Tarde's *Laws of Imitation* (1890) and later in *Social Laws* (1898). See Sampson, *Virality*.

102   Gabriel Tarde, *Monadology and Sociology* (Melbourne: Re.Press, 2012), 5. As Tarde puts it, "the hypothesis implies both the reduction of two entities, matter and mind, to a single one, such that they are merged in the latter, and at the same time a prodigious multiplication of purely mental agents in the world."

103   Ibid., 35–36.

104   Ibid., 35.

105   Ibid.

106   Ibid.

107 Ibid., 36.
108 Ibid., 8–9.
109 Ibid., 36.
110 Theo Lorenc, "Afterword: Tarde's Pansocial Ontology," in Tarde, *Monadology and Sociology*, 73.
111 Ibid., 94.
112 Didier Debaise, "The Dynamics of Possession," in *Mind That Abides: Panpsychism in the New Millennium*, ed. David Skribna (Amsterdam: John Benjamins, 2008), 226.
113 Ibid., 225.
114 Ibid.
115 Deleuze and Guattari, as cited in ibid., 226.
116 Ibid., 224, emphasis added.
117 Deleuze and Guattari, *What Is Philosophy?*, 210.
118 Ibid., 41, 109, 218.

5. RELATIONALITY, CARE, AND THE RHYTHMIC BRAIN

1 Bennett and Hacker, *History of Cognitive Neuroscience*, 113.
2 Ibid.
3 Metzinger, *Ego Tunnel*.
4 Deleuze, *The Fold*.
5 Protevi, "Deleuze and Wexler."
6 Christian Borch, "Urban Imitations: Tarde's Sociology Revisited," *Theory, Culture, and Society* 22, no. 81 (2005): 81–100.
7 Metzinger, *Ego Tunnel*, 163–83.
8 Ibid., 171.
9 Ibid., 168.
10 Ibid., 172.
11 Ibid., 105.
12 Ibid., 100–101.
13 Ibid., 101.
14 Ibid., 21.
15 Ibid., 102–3.
16 Ibid., 104–8.
17 A point made in Graham Harman, "The Problem with Metzinger," *Cosmos and History: The Journal of Natural and Social Philosophy* 7, no. 1 (2011): 31.

18 Shulman, *Brain Imaging,* 144.
19 Ibid., 78–80.
20 Ibid., 81.
21 Ibid., 78.
22 Ibid., 85.
23 Deleuze, *The Fold.*
24 Ibid., 230.
25 Ibid., 242.
26 Ibid., 231.
27 Ibid., 228.
28 Metzinger, *Ego Tunnel,* 169.
29 Ibid., 170. Metzinger makes the empathic Ego a second step in the evolution of the human preceded by the Ego Tunnel.
30 Protevi, "Deleuze and Wexler," 174.
31 Wexler, *Brain and Culture,* 22.
32 Ibid., 39.
33 Ibid., 39–40.
34 Ibid., 9. Also cited in Protevi, "Deleuze and Wexler," 173.
35 Ibid., 113–21.
36 Ibid., 115.
37 Christian Borch, "Urban Imitations," 82–83.
38 Ibid., 95.
39 Ibid., 82.
40 Ibid., 96.
41 Wexler, *Brain and Culture,* 115.
42 Ibid., 121–22.
43 Ibid., 109.
44 Ibid., 105.
45 Stiegler, *Taking Care of Youth and the Generations,* 56.
46 Wexler, *Brain and Culture,* 135–37.
47 Ibid., 81–83.
48 Ibid., 44.
49 Ibid., 80.
50 Ibid.
51 Stiegler, *Taking Care of Youth and the Generations,* 8.
52 Wexler, *Brain and Culture,* 231.
53 Stiegler, *Taking Care of Youth and the Generations,* 7–8.
54 Zoe Williams, "Is Misused Neuroscience Defining Early Years and Child Protection Policy? The Idea That a Child's Brain Is Irrevocably Shaped

in the First Three Years Increasingly Drives Government Policy on Adoption and Early Childhood Intervention. But Does the Science Stand Up to Scrutiny?" *The Guardian,* April 26, 2014, http://www.theguardian .com/education/2014/apr/26/misused-neuroscience-defining-child-pro tection-policy.

55 Ibid.

56 Ibid.

57 Jan Macvarish, author of *Biologising Parenting: Neuroscience Discourse,* cited in ibid.

58 Ibid.

59 Graham Allen and Iain Duncan Smith, "Early Intervention: Good Parents, Great Kids, Better Citizens," Centre for Social Justice, 2008, 59, http://www.centreforsocialjustice.org.uk/UserStorage/pdf/Pdf%20re ports/EarlyInterventionFirstEdition.pdf.

60 Ibid., 57.

61 Ibid., 115.

62 Wexler, *Brain and Culture,* 36.

63 Protevi, "Deleuze and Wexler," 180.

64 Ibid.

65 Ibid.

66 Ibid., 182.

67 Ibid.

68 Sampson, *Virality,* 122.

69 P. Fritzschep, *Life and Death in the Third Reich* (Cambridge, Mass.: Harvard University Press, 2009), 61.

70 Protevi, "Deleuze and Wexler," 182.

71 Obsolete Capitalism, *The Birth of Digital Capitalism: Crowd, Power, and Postdemocracy in the Twenty-First-Century* (Obsolete Capitalism Free Press, 2015), v, http://issuu.com/obsoletecapitalism/docs/the_birth_of _digital_populism_for_i.

72 BBC *Sunday Politics,* October 19, 2014, the Canvey Island Independent Party shows support for the UKIP.

73 Reich, in Protevi, "Deleuze and Wexler," 178.

74 Foucault's preface to Deleuze and Guattari, *Anti-Oedipus.*

75 Reich, *Mass Psychology of Fascism,* 55.

76 Obsolete Capitalism, *Birth of Digital Capitalism,* xx–xxxvii.

77 It must be noted that the UKIP leader resigned after narrowly failing to win a seat in Parliament in the May 7 General Election of 2015. This

was despite a significant swell in support for UKIP, amounting to almost 13 percent of the national vote share. Nonetheless, by May 11, he was back as leader and began a purge of his critics inside the party.

78 Tarde, *Monadology and Sociology*, 36.

79 Reich, *Mass Psychology of Fascism*, vii.

80 Wexler, *Brain and Culture*, 83.

81 Protevi, "Deleuze and Wexler," 230–31.

82 Wexler, *Brain and Culture*, 231.

83 Zajonc, as cited in ibid., 155–56.

84 "FDA Permits Marketing of First Brain Wave Test to Help Assess Children and Teens for ADHD," press release, http://www.fda.gov/news events/newsroom/pressannouncements/ucm360811.htm.

85 David Grant, *That's the Way I Think: Dyslexia, Dyspraxia, and ADHD Explained* (Oxon, U.K.: Routledge, 2010), 157.

86 Borch, "Urban Imitations"; Thompson, "Time, Work-Discipline, and Industrial Capitalism"; and Virilio, *Speed and Politics*.

87 Stiegler, *Taking Care of Youth and the Generations*, 99–100.

88 Judith Becker, "Rhythmic Entrainment and Evolution," in Berger and Turow, *Music, Science, and the Rhythmic Brain*, 67–68.

89 Barbara Ehrenreich, *Dancing in the Streets: A History of Collective Joy* (London: Granta, 2007), 152.

CODA

1 William R. Uttal, *Neural Theories of the Mind: Why the Mind Brain Problem May Never Be Solved* (London: Routledge, 2014), 260–62.

2 Metzinger, *Ego Tunnel*, 220.

3 Ibid., 238.

4 Ibid., 222.

5 Ibid., 233–37.

6 Huxley, *Doors of Perception*, 6.

7 Ibid.

8 Ibid.

9 Thomas De Quinicey, *Suspriria De Profundis: Being a Sequel to Confession of an English Opium Eater* (repr., Edinburgh: Adam and Charles Black, 1884), viii.

10 Ibid., 3.

11   Huxley, *Doors of Perception,* 14.
12   Ibid.
13   Ibid.
14   Ibid.
15   Ibid., 4.
16   Ibid.
17   Ibid., 4–5.
18   Ibid., 5.
19   Ibid.

Index

Tony D. Sampson is reader in digital culture and communications at the University of East London. His publications include *The Spam Book* (coedited with Jussi Parikka) and *Virality: Contagion Theory in the Age of Networks* (Minnesota, 2012). He is a cofounder of Club Critical Theory: Southend and director of the EmotionUX Lab at the University of East London.